Managing the Urgent and Unexpected

Managing the Urgent and Unexpected

Twelve Project Cases and a Commentary

STEPHEN WEARNE AND
KEITH WHITE-HUNT

 Routledge
Taylor & Francis Group

LONDON AND NEW YORK

First published 2014 by Gower Publishing

2 Park Square, Milton Park, Abingdon, Oxfordshire OX14 4RN
52 Vanderbilt Avenue, New York, NY 10017

Routledge is an imprint of the Taylor & Francis Group, an informa business

First issued in paperback 2020

British Library Cataloguing in Publication Data
A catalogue record for this book is available from the British Library

Library of Congress Cataloging-in-Publication Data
Wearne, S. H.
  Managing the urgent and unexpected : twelve project cases and a commentary / by Stephen Wearne and Keith White-Hunt.
      pages cm. -- (Advances in project management.)
  Includes bibliographical references and index.
    ISBN 978-1-4724-4250-5 (hardback) -- ISBN 978-1-4724-4252-9 (ebook) 1. Construction projects--Management--Case studies. 2. Emergency management--Case studies. 3. Public works--Management--Case studies. 4. Project management--Case studies. I. White-Hunt, Keith. II. Title.
  TA190.W39 2014
  624.068'4--dc23

                                                                    2014010829

ISBN 13: 978-1-4724-4250-5 (hbk)
ISBN 13: 978-0-367-67008-5 (pbk)

Contents

List of Figures

List of Tables

Preface

Urgent work can be required unexpectedly to take advantage of a business opportunity, or for protection against an unexpected sudden physical or business threat, or to restore a severely damaged asset. Any such urgent and unexpected work demands an instant start, in contrast to the often lengthy processes of investigation, evaluation, development, selection and planning normal in businesses and public services before any proposed work is started.

This book explores what is different managerially if work is unexpected, its implementation is urgent and an immediate start it is required. The book draws on twelve cases ranging from the launch of the Freeview television system in the United Kingdom to the sifting and removal of the New York World Trade Center pile of debris following the 9/11 terrorist attack. The book summarizes how they were managed, demonstrates that opportunities may sometimes be created in the face of adversity and suggests how normal organizations can prepare to manage abnormal demands.

Urgent and unexpected projects have to be rare in business or government to be economically and socially tolerable. The chance that any one person except those in the emergency services will ever manage such a project is small. It's not possible to know who should learn these lessons in order to be prepared. It's their organizations that can and should be prepared. The lessons offered here should help private and public organizations plan how they should authorize and support future urgent work if unexpectedly needed to take advantage of new business opportunities or to protect or restore systems and services.

Acknowledgements

Thanks are due to companies and public services for access to cases, to many individuals for information and comments, to postgraduate students for their thoughts and additional sources, and to UK and US authorities and the Major Projects Association for invitations to conferences.

PART I
Narrative

Chapter 1
You Get a Start!

What's Different?

Wise managers think ahead. Whether proposing a new product or process, a change of management or the purchase of a new building or IT system, the project should start with an objective agreed with committed stakeholders, an unambiguous definition, scope, plan, risk assessment, budget accepted by all parties, and prepared resources led by an empowered individual and project team supported by a trusted information system. In their own words this is what managers say when reflecting on their experience. In effect so say many books, textbooks, government reports and company guides in stating what is required to have the best chance of successful results.

So what is different if a project is urgent and unexpected given that it is to be started without prior study of its scope or how to deliver it? How is managing any urgent and unexpected project different from managing normal work? What are the lessons and ideas which may help? The purpose of this small book is to review some real-life urgent and unexpected projects and how they were started and managed to try to find common answers to these questions.

Twelve cases of urgent unexpected projects provide the basis of this book, drawing on reports and observations of how they were organized and managed. Use is also made of the more limited information available on some other urgent and unexpected projects. All these projects varied in their size, nature, resources then available for immediate use, the types of organizations involved, local conditions and public concern. All used engineering resources. All achieved their sponsor's objectives. They do not necessarily represent every possible surprise which may require immediate action. They do provide an opportunity to consider what was different in managing them compared to normal practice.

The Cases

In two of the twelve cases the urgent work was required to respond to unexpected opportunities to offer new services (see Table 1.1).

Table 1.1   Urgent work in response to unexpected opportunities

Short name and location	Scope
New TV business, United Kingdom	Design and launch the UK Freeview digital television system in six months, requiring an immediate start to technical development and consumer marketing to meet the criteria of the licensing authority and provide the basis for the successful launch and operation of the system.
Temporary rail station, United Kingdom	Construction of a temporary new rail station including platforms, footbridge, waiting room and car park to enable trains to connect both sides of a town after the collapse or closure of the town's road bridges.

In three cases urgent unexpected engineering work was required for the protection of urban areas against physical threats (see Table 1.2).

Table 1.2   Urgent unexpected work to save assets under threat

Short name and location	Scope
Thames bank raising, London, United Kingdom	Raise and strengthen 38 miles of the Thames tidal river banks and linked structures within six months. The construction work to be done had to suit the state of the banks and other defence structures in more than 800 properties.
Flood diversion scheme, Chichester, United Kingdom	Design and construct in two weeks an emergency river flood diversion system of pumping, piping, temporary channels and culverts around the city.
Ouse banks raising, Yorkshire, United Kingdom	Urgent replacement of emergency sandbagging with sheet piling along long lengths of the river banks after heavy flood flow when the bank was waterlogged and overtopping.

Two cases were work to save threatened sections of assets (see Table 1.3).

Table 1.3   Urgent unexpected work to save assets under threat

Short name and location	Scope
Propping of motorway viaduct, London area, United Kingdom	Support temporarily fire-damaged pre-stressed concrete main beams of the M1 motorway by props, spreaders and smaller props in order to permit partial resumption of motorway traffic the next day.
Embankment stabilization, Heck, Yorkshire, United Kingdom	Stabilize a 1.8 km stretch of the East Coast Main Line railway embankment de-stabilized by flood water.

In four of the twelve cases the urgent unexpected work was required to restore failed or severely damaged assets (see Table 1.4).

Table 1.4   Urgent unexpected work to restore failed or very damaged assets

Short name and location	Scope
Remote bridge repair, Northern Australia	Make and install a temporary deck structure spanning the ruptured central spans of a remote major highway bridge over a wide crocodile-infested tidal river, as the urgent action pending pier repair and permanent replacement spans. Plus some upgrading of other bridge piers.
Temporary power line, Auckland, New Zealand	Design and construct a temporary emergency electricity 160 MVA 110V transmission line 9.8 km long and connecting work to restore the power supply to the Auckland Central Business District utilizing a railway tunnel and open-wire transmission.
Aire banks restoration, Yorkshire, United Kingdom	Temporary raising of a section of the River Aire banks, emergency repairs and strengthening of flood-retaining embankments and construction of evacuation sluices and outfalls.
Rail reinstatement, Great Heck, Yorkshire, United Kingdom	Restore main line tracks, overhead power and the signalling systems over 1 km of the East Coast Main Line. The damage extended to both rail tracks, the overhead power systems, ground cables and signalling, over a length of 1 km of track, and minor damage to a road overbridge and farming property.

And one case was massive work with the principal objective of finding survivors and evidence of the remains of buried victims of terrorism (see Table 1.5).

Table 1.5   Urgent unexpected work to find survivors and recover evidence of victims

Short name and location	Scope
9/11 pile sifting, New York, USA	Sift, make safe and remove 1.6 million tons of rubble, hazardous major structural elements and other wreckage ('the pile') in order to search for survivors and remains and clear the 9/11 site in Manhattan, over 10½ months. The work was urgent initially to search for survivors, and then find all identifiable remains whilst clearing the site.

The sponsors of these urgent and unexpected projects were the owners and operators of the assets.

Appendix 1 gives a summary of each of these cases and lists the sources used. The more limited information from other urgent and unexpected work is listed in Appendix 2.

Projects started urgently in order to meet a date set by a licence, financial support or a political promise but for which there was no intention to deliver the project any faster than normal after their start have not been used for this book as they do not meet the criteria of being both urgent and unexpected.

Other surprise events such as failures of products, services and projects are reported in the media and company or government publications. Many of these reports indicate whether the results of actions taken were successful, unsuccessful or in between. Unfortunately for our purposes of learning from them they do not include information on how the urgent unexpected work required was managed.

Limitations

There are some limitations to studying how urgent and unexpected projects are managed in comparison to normal practice.

One limitation is that arrangements to observe unexpected events cannot be made in advance. Learning from the unexpected is inevitably retrospective and its scope limited by the people and other sources of information accessible after the event. Published reports, conference papers, some interviews and individuals' comments provided the information on the cases that are the basis of this book.

A second limitation is that among the many publications on the management of projects there are few studies that show what normal practice is. Since the classic first article by Gaddis (1959) on his experience and the pioneering analysis of some major construction projects by Morris and Hough (1987), many authors have offered concepts, lessons from case studies and techniques that seem logical for managing projects. Causes of success have been stated from various studies, but with little independent evidence showing the results of their testing on further projects. Causes of failure have been stated from such studies, but a model based upon avoiding causes of failure doesn't establish what is normal.

Most of the principles and techniques for managing projects to be found in books and in company manuals may thus only be as valid as the common belief in Europe 600 years ago that the world is flat. At best they are a logical guide to thoughtful small steps beyond experience. They provide the advice summarized at the start of this chapter that a project should normally start with an objective agreed with committed stakeholders, an unambiguous definition, scope, plan, risk assessment and budget accepted by all parties, and prepared resources led by an empowered individual or project team supported by a trusted information system. What managers and analysts say is good advice is not always followed in practice, but this advice provides a model for comparison with projects that are not normal because they are urgent and unexpected.

Emergencies and Crises?

An emergency is defined as a sudden event requiring immediate action where there is thought to be a threat to life or property. None of the cases were emergencies in that sense, but four of them consisted of managing restoration and recovery work after emergencies. Four consisted of managing preventative work to forestall threatened emergencies. Two were new business opportunities. These two and the other three cases are apparently rather different. In their effects all twelve cases are the same in being triggered by a surprise event which demanded resources for urgent action.

If a project is unexpected and also urgent the result can be a crisis, what the dictionary defines as 'a turning-point in progress' or 'a state of affairs in which a decisive change for better or worse is imminent' (*Oxford English Dictionary*) and medics say 'when a change takes place which is decisive of recovery or death'. British Standard PAS 200: 2011 (BSI 2011) provides guidance on anticipating and managing business crises. It is similar to the advice on managing a project. The word 'crisis' implies that the result will be bad, or at least undesired. This is not necessarily true. Two of the cases examined show that unexpected events can be opportunities. The crisis then can be in applying resources quickly enough.

Every Case is Unique

Every project tends to seem unique to the individuals and organizations involved in it. An urgent and unexpected project may seem to those involved in it to be more unique than others. In this book we look for similarities and for differences in how twelve urgent unexpected projects were managed. To do this, Chapter 4 reviews how they were started. Chapter 5 reviews how they were managed. Chapter 6 then draws on these to compare reviews of urgent and unexpected projects with normal ones. Chapter 7 is a check list with comments on how business and government organizations can prepare for future urgent unexpected work. Chapter 8 summarizes what's different in managing urgent unexpected projects and also lessons for 'normal' projects. But first, Chapters 2 and 3 first explore what is meant by describing work as 'urgent' and as 'unexpected'.

Chapter 2
Urgent?

What is Urgent?

The dictionaries tell us that 'urgent' means demanding prompt action, priority, of pressing importance and working faster than normal.

Though many people might agree with this definition, they might differ in what they mean when they say something is urgent, or in what they understand when others use this word. What is seen as urgent at one level in an organization may not be seen as urgent at other levels, as noted by McDonough and Pearson (1993) in studying the impact of urgency on the performance of product development work. Various stakeholders, suppliers and other contributors to a project could similarly vary in their understanding and their response to work stated as being urgent. As was observed in another large company, working faster than normal may or may not mean priority in the use of resources. Interpretations of 'urgent' varied from a very strong and commanding 'do it now at the expense of everything else' to 'business as usual – only quicker'.

In the way the word 'urgent' is used in practice there can thus be degrees of urgency which are open to different understandings. But they all agree that it means working faster than normal, and that this usually incurs increased cost.

Cost, Time and Performance

It is normal for all organizations to plan their work to try to use resources continuously, to avoid the cost of employing people and material resources unproductively. An unplanned temporary use of more resources to work faster for one project defined as urgent may thus incur greater costs. Alternatively, pressure to avoid these greater costs may cause reduction in the quality or safety of the work.

This potential conflict in priorities between the time allowed to deliver some work, its cost and its quality was depicted by Martin Barnes in his diagram now known as the 'iron triangle', shown in Figures 2.1 and 2.2. In this the term 'performance' embraces the quality, safety and fitness for purpose of the work done. The triangle indicates that priority to performance, to cost or to time is possible only at the expense of the other two.

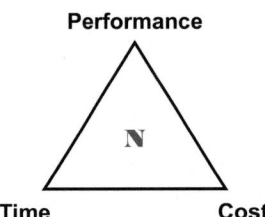

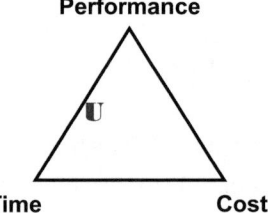

2.1   Cost–time–performance triangle (Barnes 1971)

2.2   Urgent project objectives

The position of the letter **N** in Figure 2.1 indicates work for which these three variables are considered to be equally important. Figure 2.2 illustrates how the balance is different if the work **U** is urgent so that performance and delivery objectives override cost.

The iron triangle is a simple illustration of what can be complex decisions between qualitative and quantitative objectives, but using it to display what is thought to be the relative importance of the three variables may be helpful in checking whether all parties agree on what has priority. It can also remind them throughout the work of their commitment to the urgency it indicates.

Economic Speed of Projects

Agreeing that a piece of work is urgent doesn't answer the question of how much extra cost to incur by working faster than normal.

One choice is to plan to work at the speed which incurs the minimum cost. Or, alternatively, to work at maximum speed. Or at a speed between the two extremes. Estimates of the likely cost of the work over a range of possible speeds of work are needed to decide between these choices and to plan the total time to be allowed for its delivery. Given this data the speed should be selected at which the net present value of extra financial return obtained earlier by faster working balances its extra cost. The same criterion should be used whether once delivered the product is expected to earn a financial return or whether to provide a social benefit that is measured in financial terms.

Figure 2.3 illustrates how the speed planned for delivering work affects its cost. Some costs increase with the time taken, as indicated by the C_t curve in the diagram, for instance financing the use of resources. Other costs decrease, as indicated by the C_m curve, for instance the direct costs of hired resources. The sum of the two is shown by the Total Cost curve.

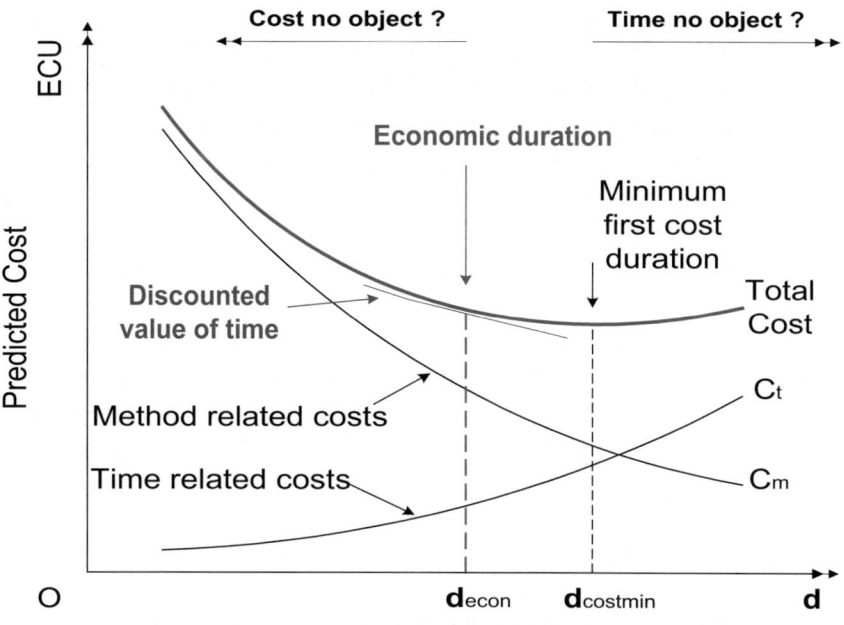

SELECTED DURATION OF THE PROJECT

2.3   Relationship between planned duration and predicted cost of delivery

Continuous curves are a simplification in Figure 2.3. In practice, the relationships may include step changes in choices in the capacity of resources.

In principle the relationship indicated in Figure 2.3 between planned duration and consequent cost provides a basis for classifying work required into three distinct degrees of urgency:

- Minimum initial cost – If the objective is to complete for minimum cost, its planned duration should be where Total Cost is a minimum, the $d_{costmin}$ point shown in the diagram. This is the condition for investment in services which will not earn or save money or the benefits they produce will not be credited financially. Urgency in terms of earning a financial return is then zero.

- Economic duration – If the objective is to produce goods or services which are expected to earn a financial return, greater expenditure than minimum cost is logical to try to achieve delivery earlier than $d_{costmin}$. This point on the Total Cost line is where its slope (shown by a tangent line) represents the discounted amount which is expected to be earned per week after delivery (the slope of this line is negative, as it represents not cost but financial return per unit of time). On this basis d_{econ} should be chosen as the planned duration of the work. It is the commercial condition 'Time is money'.

- Minimum time – If speed completely overrides the criterion of cost, any attention to optimizing the use of resources is irrelevant. Time is priceless. Only physical conditions, resources and safety limit the speed of work.

Normal Time or Natural Pace

As discussed above, work which has the objective of providing or improving services which will not earn money or not be given financial credit should be planned to be delivered in time $d_{costmin}$, as shown in Figure 2.3. In commenting on this from experience and studies in construction, Pilcher (1967) called this the 'normal time'. It indicates the financially logical speed of work for a contractor or other supplier who will be paid a fixed amount for delivering a package of work. Others call this speed of work the 'natural pace' for using resources most economically. Planning to work at this speed so as to incur least cost is financially sensible for any organization due be paid a fixed price for delivering goods or services, whether or not the customer wants them faster.

So What is 'Urgent'?

Logically, work which has the objective of providing or improving services which will not earn money or be given equivalent financial credit should be planned to be delivered in time $d_{costmin}$. But if the work is also stated to be urgent, this must mean that there is an overriding reason for deciding to use resources uneconomically so as to try to deliver it earlier than in time $d_{costmin}$. Or work to produce goods or services which are expected to earn money should be planned to be delivered in time d_{econ}. If such work is also stated to be urgent, this must mean that there is an overriding reason for deciding to use resources uneconomically so as to try to deliver it earlier than in time d_{econ}.

The reasons why the twelve cases listed in Appendix 1 were stated to be urgent are summarized in Table 2.1.

Table 2.1   The twelve cases: Reasons for urgency

Case	Why urgent
New TV business	To bid for the licence and be ready enter into the digital TV market by the time limits set by the licensing authority.
Temporary rail station	To provide railway service to reconnect two sides of town after loss of road bridges.
Thames bank raising	To protect much of London from higher water levels that might occur before provision of a 'permanent' Thames tidal barrier.
Flood diversion	To prevent predicted imminent high river flow from repeating city flooding in previous years.
Ouse bank raising	To replace potentially unstable emergency protection of housing and farms.
Viaduct propping	To make the viaduct safe for a peak of traffic expected the next evening.
Embankment stabilization	To minimize loss of business and public rail service.
Remote bridge repair	To restore the only all-weather highway link into a large industrial, defence and tourist region as soon as possible.
Temporary power line	To minimize the length of the period of loss of power to city business district and public services.
Aire banks restoration	To prevent recurrence of flooding and allow return to village and restoration of houses, farms and community.
Railway reinstatement	To minimize loss of business and public rail service, plus political pressure to restore service following previous but unrelated losses of services.
9/11 pile removal	Automatic individual and collective response to try to rescue survivors and search for identifiable human remains.

The importance financially and socially of these public and business services was the common reason for using resources as fast as possible in most of the cases. Many of these assets are part of complex and interdependent systems which tend to be used up to their maximum capacity. Their daily value when in use is overwhelmingly greater than the costs of working as fast as possible within the limits of physical conditions, safety, resources and logistics. In the new TV business case the effect of the value of meeting the competitive deadline date set by the regulating authority was the same. Speed in sifting and removing the 9/11 pile was at first thought to be literally vital, but urgency continued to be dominant in the search for identifiable remains. This last case is the extreme in scale and communal impact amongst the twelve, but though the triggering events causing urgency thus varied greatly from case to case, the common factor in all these projects was that the cost of working as fast as possible was not a factor in deciding to start them.

In the bridge repair and railway reinstatement cases the work had to be substantially complete for safe restoration of those public services. In the Thames bank raising case, urgency was sustained because flood walls are no protection until complete. The flood recovery cases were the same. In the new TV business case, until the deadline date; in the 9/11 pile case, until the last chance of identifying traces of the many unaccounted-for casualties. Logistics were the common restraint.

Other Urgent and Unexpected Projects

Working as fast as possible within the limits of physical conditions, safety and resources is similarly seen in the notes of the other urgent and unexpected projects listed in Appendix 2. For instance, the massive and lengthy reinstatement of public services after the Japanese Great Hanshin-Awaji (Kobe) and the US Northridge earthquakes, in Carlisle after flooding of the city centre, and in East Timor after deliberate destruction. As in Auckland, restoration of public power systems was equally urgent in Darwin after widespread cyclone damage. In the case of the repair of the Webbers Falls Bridge in Ohio, priority in restoring services to the travelling public and local stakeholders was similar to the decisions on the urgent repair of the remote highway bridge in Australia. Many bridge repair projects, whether in urban, farming or industrial areas, are similarly urgent, such as the viaduct repairs example listed in Appendix 2. And, for the same reason, the Hatfield railway and New York subway restoration work. Commercial loss was a clear reason for urgency in restoring the Gulf of Mexico Thunder Horse oil well drilling riser, and environmental protection for the UK case of carcass disposal after an animal disease epidemic.

Urgency Cost

In some of the twelve cases and the other urgent and unexpected projects estimates of possible cost were requested to advise stakeholders and to arrange for payment for the work done, but predictions of cost were not a criterion for deciding to initiate the work. This distinguishes them from projects in businesses and public service which are 'normal' in the sense of being systematically investigated, evaluated, planned and budgeted before starting them. That is not to say that the twelve cases and these other urgent and unexpected projects were otherwise similar to each other. They varied in size, duration, technology, novelty and context and they probably felt unique to every person involved in them. But they had in common that they were unexpected and that they were urgent compared to normal projects in their organizations and industries. In all of them, the value of delivering the work to be done was agreed to be overwhelmingly greater than the likely extra cost of working as fast as possible. This gave the word 'urgent' a clear meaning.

Chapter 3
Unexpected?

What is Unexpected?

Unexpected is defined in dictionaries as meaning an event 'not regarded as about to happen'. That is a clear definition. In practice the word has a range of meanings. For instance, unexpected can mean that an event was never thought of previously. Or it can mean that the event was thought of but the possibility of it happening was not allowed for. So knowing what was really meant when it is stated that a project was unexpected may be needed to help understand how that project was planned and managed.

Following the aircraft and process industries it is now common to classify risks to projects as either 'unknown unknowns' or 'known unknowns'. Under this classification the causes of events not thought of are 'unknown unknowns', and the causes of events that were thought of are 'known unknowns'. Unknown unknowns are sometimes called 'totally unknown', so emphasizing that they were never thought of. Or 'unknown in *nature*', indicating that this type of event was never previously thought of.

Known unknowns can be subclassified. In commenting on lessons from emergency projects, Dalton (2003) suggested two categories of choices in preparing to avoid or minimize the risk of hitherto inconceivable events. In the first category may be events for which no reasonable person would prepare. They are uninsurable because probabilities cannot be ascribed to the inconceivable. In the second category are those which, if they do occur, should have been anticipated notwithstanding how improbable the eventuality. This classification gives us two classes of known unknowns. The difference may seem irrelevant in managing actions to remedy the consequences, but could affect preparedness and restraints on these actions.

Other terms are also in use, for instance calling unknown unknowns 'black swans' to indicate that they are expected to be rare (Prieto 2011). Using the word rare implies that the event may have been thought of but disregarded as improbable. More helpful in understanding how such projects were planned and managed may be a set of classifications suggested by Geraldi et al. (2010) when studying project managers' responses to unexpected events that affected their projects. They classified those events according to their *probability, impact, pertinence* and *timing. Probability* and *timing* seem to be clear and understandable classifications of known unknowns. By the word 'impact' Geraldi and colleagues did not necessarily mean impact in the physical sense, so the word *scale* is perhaps better to use with *nature, probability* and *timing* to classify unexpectedness.

The Case Studies

All the triggering events in the twelve cases were *pertinent* in the sense used by Geraldi and colleagues when studying project managers' responses to unexpected events; that is, they were the causes of these urgent and unexpected projects. The cases varied in why they were not regarded as about to happen, as summarized in Table 3.1.

The opportunity for the new TV business was unexpected in its *probability*. The industry was aware that the joint venture which had won the licence to launch a digital TV system might be in difficulty financially, but did not expect it would fail. The team formed for the project had no previous experience of working together. For them it was unexpected in its *nature*.

The concept of providing a temporary rail station to help restore links between two parts of a town was unexpected in its *nature*. The project team commonly used scaffolding during maintenance and improvement work, and at short notice. The material and skills required to create the temporary rail station were unexpected in *scale*.

The risk of a storm surge threatening the Thames estuary with flooding was a known unknown. Recognition that interim protection of London from higher water levels should not wait six years for the 'permanent' barrier required flood protection systems earlier than expected, so this can be classified as unexpected in its *timing*. The consequent temporary bank raising project was unexpected in its *nature*. Similarly, the risk of the river flooding in Chichester was a known unknown, but the peak predicted was unexpected in its *timing*. All parties had agreed on the need for additional flood diversion capacity, but sustained extreme weather conditions indicated the need to proceed immediately with the emergency project. The consequence was a temporary work project unexpected in its *nature*.

The risk of flooding was a known unknown for the Aire and Ouse rivers, but sustained heavy rainfall and low atmospheric pressure had caused these tidal rivers to rise above their highest ever recorded levels. In these cases the extent of flood damage which occurred was unexpected in its *scale*, and the consequent project work was unexpected in its *scale*. Flooding threatening the stability of the Heck railway embankment was an unknown unknown, unexpected in its *probability*. The consequent project of waterlogged embankment maintenance was unexpected in its *scale*.

The weakening of a road traffic viaduct by an industrial waste fire was unexpected in its *nature*. The amount of propping was greatly in excess of the temporary support occasionally required for maintenance and repairs, so this project can be classified as unexpected in its *scale*. Similarly, train derailment requiring repair of track and services is a known risk of railway systems, but not as a result of the massive damage initiated by a road vehicle precipitating the derailment and collision of trains on the scale that occurred at Great Heck. Its cause can thus be classed as an unknown unknown, unexpected in its *nature*. The scale of reconstruction required after this major crash was greatly in excess of the work expected in the regional provision of resources for maintenance and repairs, and so this project can be classified as unexpected in its *scale*.

In the case of the repair of a remote bridge, corrosion in the pier steel was a known unknown, but it occurred seven times faster than expected in the given conditions. Applying the classifications of unexpectedness discussed above, the event which triggered the project can be classed as a known unknown, but unexpected in its *timing*. The consequence was much heavier engineering than the repair work expected in bridge maintenance, and so the urgent work for this project can be classified as unexpected in *scale*.

By contrast, in the temporary power line case, the overheating failure of the oil-filled power cables in their buried conditions was an unknown unknown. This event had not been thought of at all, by any party. No such previous fault had been reported, so this can be classified as unexpected in its *nature*. Undertaking to design, plan and supply the consequent temporary power line and manage agreements with third parties was unusual for the project team and so may also be classified as unexpected in its *nature*.

In the case the 9/11 pile removal, the possible effects of an aircraft accidentally hitting a tower had been considered, but not the consequent destructive effects of fire from aviation fuel. The triggering event had been envisaged, but not the result. Until it happened, the asset owners, their insurers and other stakeholders would perhaps have said that investing in protection against such an event would be absurd. That cause of this event can therefore also be classified as an unknown unknown, unexpected in its *nature*. The scale of the consequent rubble sifting and searching work was overwhelmingly beyond the experience of all parties and the search-and-rescue capacity of the emergency services and the infrastructure maintenance services, so that the project can be classified as unexpected in its *scale*.

Table 3.1   The twelve cases: Unexpectedness

Case	Triggering event
New TV business	Unexpected in *probability*
Temporary rail station	Unexpected in *nature*
Thames bank raising	Unexpected in *timing*
Flood diversion	Unexpected in *timing*
Ouse bank raising	Unexpected in *scale*
Viaduct propping	Unexpected in *nature*
Embankment stabilization	Unexpected in *probability*
Remote bridge repair	Unexpected in *timing*
Temporary power line	Unexpected in *nature*
Aire banks restoration	Unexpected in *scale*
Railway reinstatement	Unexpected in *nature*
9/11 pile removal	Unexpected in *nature*

Though the twelve cases thus varied in their unexpectedness, some unexpected in *probability* or in *nature*, others in *scale* or in *timing*, all were unexpected in the sense that the results of the triggering events demanded urgent action and a scale of work not envisaged in planning the provision of resources for improvements, safety, maintenance and recovery of the established assets and services (Table 3.1).

Other Urgent Unexpected Projects

The same classifications can be applied to the other urgent unexpected projects listed in Appendix 2. The Carlisle and the Daly River flooding, the Darwin cyclone destruction, the Kobe and the Northridge earthquakes, the Manchester bomb, the New York subway crash and the Webbers Fall bridge damage were all extreme events unexpected in *scale*. The animal epidemic support work was unexpected in *probability*. The Hatfield rail failure was unexpected in *timing*. The East Timor sabotage, Heathrow tunnels collapse and motorway viaduct weld cracking were unexpected in *nature*.

Unexpectedness and Urgency

Whether the events which triggered any of the twelve cases should have been expected and resources prepared for them is a separate question. In some cases the events led to reconsideration of what resources and other preparations should in future be available at short notice for that sort of event. At the time they caused unexpected work. It is this and their urgency which conditioned the start of these projects.

Chapter 4

Project Starts

Project Sponsors and Stakeholders

The advice in textbooks and other guides that a project should start with agreement between the sponsor and other stakeholders on its objectives provides a basis for reviewing the start of each of the twelve cases.

In the TV business opportunity the UK's national radio and television broadcasting corporation was the sponsor of the project. The other stakeholders in the project were two companies with complementary businesses. The ultimate stakeholders were the future viewers as represented by the regulating authority and the UK government in their policy of encouraging the change from analogue to digital television nationally. The three businesses worked together for the agreed objective of designing and creating a new asset project under the time limit and other terms set by the public broadcasting regulating authority.

The owners of the track were the sponsors of the temporary Workington North rail station, working closely with the train operating companies who were familiar stakeholders but for this project also with local and central government departments and the flood emergency recovery group.

In the Thames interim bank raising case the need for some form of new flood protection was already agreed by the sponsor, the urban authority. Consultative procedures established between the urban authority and the central government provided the basis for agreement on proceeding with the interim protection project. The interim project required the addition of temporary teams located along the length of each bank to consult the many riverside property owners and stakeholders and then plan and manage mostly small-scale simultaneous operations in these many stakeholders' properties.

The national authority responsible for major river safety and management resources and the local authority were similarly the sponsors in the flood diversion case. The need for the urgent and unexpected project was agreed by the sponsors and other stakeholders through emergency planning meetings which had been the forum for agreement on the longer-term project. Local consultation was added with the property owner and occupier stakeholders affected by the temporary urgent and unexpected project.

The Public Works Department was the sponsor of the remote bridge repair project. The many other stakeholders were the users of the highway – business, private and government. Urgency in repairing the bridge was not diluted by the sponsor's other immediate tasks of arranging a temporary ferry service and substitute road routes. The decision to give priority to restoring the crossing was undisputed. There were no local stakeholders in the bridge area who might have sought additional objectives with local advantages such as the development of a community around a project for a replacement crossing or for security and maintenance.

In the case of the temporary power line, the private utility company similarly accepted the role of sponsor. Their immediate tasks had been to arrange and coordinate local standby and other emergency power supplies. The major public stakeholders were the city businesses and domestic power consumers. The urgent unexpected project was proposed and managed by the contractor's team. Central and local government leaders facilitated consent to the project under emergency legislation. Cooperation followed from common agreement on the priorities for temporary restoration of power supplies.

The national authority responsible for major river safety and management was the sponsor of the river bank flood repair projects. The many local householders and businesses in these populated areas were the many public stakeholders. As is generally the case in the United Kingdom, meetings to plan for emergencies were already held regularly between the sponsor, local government, the emergency services, police and business representatives. These meetings established the relationships for consultation and agreements needed on the sponsors' decisions to start the flood damage repair projects, but in each case with the addition of detailed consultation with the property owners and occupiers and other stakeholders affected. With central government support the sponsors agreed that the urgent protective work should include whatever improvements to protection could be included in the time available for its completion.

In the railway reinstatement case the track operating company was sponsor of the urgent and unexpected project. The train operating companies were the other direct stakeholders, and through them the travelling public and the freight transport companies. Reinstatement of the asset to its previous specifications was agreed immediately between the sponsor and the safety authorities, as all the track and systems had previously been up to the latest standards and changes of specification which would have otherwise required time for submission for safety approval were therefore not required. No fault in the track or systems had caused the precipitating events or the resulting damage. The local emergency planning meetings established between the asset owners, local government, the police and the emergency services provided the basis for coordination around the work.

The physically and organizationally simpler case of the stabilization of the Heck railway embankment was otherwise similar, with the track operating company the sponsor and proceeding subject to agreements on access and approval by the safety authorities.

In contrast to the above cases, for the 9/11 pile removal case the users and owners of the World Trade Center buildings, their employees, the fire and police departments, city inhabitants, businesses and government of the whole city and beyond were all, in effect, stakeholders. The fire service, police and others on site began their immediate rescue work independently and to some extent in conflict in the chaotic, crowded and risky conditions. As a result of the leadership provided by their expert staff on site, the city authorities became the sponsor of the pile removal, clearance and removal project, together with being responsible for attention to the safety of the site, lost services and damaged adjacent structures.

This variety of roles and relationships between the sponsors and the other stakeholders in the twelve cases is summarized in Table 4.1.

Table 4.1   The twelve cases: Sponsors and stakeholders

Case	Sponsor's role in the urgent unexpected project	Relationship with the other stakeholders
New TV business	Public broadcasting corporation sponsored the project and cost shared by consortium	Sponsor was responsible for relationships with the licensing authority and regulator
Temporary rail station	Rail operating company sponsored the work and met cost as a contribution to regional recovery	Sponsor responsible for relationships with local government and train operating companies
Thames bank raising	Greater London government sponsored the project; costs shared with central government	Sponsor responsible for all relationships with government and the public
Flood diversion	Environment agency and local government sponsored the project; costs shared with central government	Sponsors responsible for all relationships with government and the public
Ouse bank raising	Public environment agency sponsored the project; costs met by local and central government	Sponsor responsible for all relationships with government and the public
Viaduct propping	Consortium responsible for the management and maintenance of the system and linked roads were sponsor	Sponsor responsible for relationships with police and public
Embankment stabilization	The railway owner sponsored the project and met the cost	Sponsor responsible for all relationships with government and the public
Remote bridge repair	Territory highways department sponsored the project and met the cost	Sponsor responsible for all relationships with government and the public
Temporary power line	Utility sponsored the contractor and met the cost	The utility responsible for all relationships with government and the public
Aire banks restoration	Public environment agency sponsored the project. Costs met by local and central government	Sponsor responsible for all relationships with government and the public
Railway reinstatement	The railway owner sponsored the project. Costs met by vehicle insurer	Sponsor responsible for all relationships with government, train operating companies and the public
9/11 pile removal	City's Design and Construction Department took on sponsor's coordinating role and met costs	Mayor took on relationships with the public, state and federal governments

Formation of Project Teams

In the twelve cases the goal was well defined but the method for achieving it was not. Neither was a project team already established. As observed by Turner and Cochrane (1993), the start of a project is inevitably concentrated on defining the scope of work and the mode of operation of the project team.

Temporary structures for project management were established in all twelve cases, in different forms. In none of the sponsoring organizations were there established roles or procedures for managing these urgent and unexpected projects. Some relevant experience was available in two cases. All the project teams drew on specialist advisers. Apart from some specialist advisers all the members of the temporary teams were dedicated full-time to these projects.

For the new TV business opportunity the sponsor appointed a project manager with experience of coordinating internal projects, supported by a small steering team. Staff of the three organizations formed a virtual team for the project, working in their parent offices, with one office being 60 miles away from the others.

In the cases of the work needed after large-scale flood damage the sponsors appointed project managers and supporting staff drawn from their teams already established for impending improvement projects as well as maintenance work. In the two cases of urgent flood protection the sponsors formed project management teams led by staff established for the longer-term protection projects already planned, augmented by consultants' staff for these urgent and unexpected projects.

For the brief urgent viaduct propping the operations director led an incident management and recovery team of functional heads to plan and direct the project full-time over the days. Similarly, the head of property services and the senior engineering and construction staff formed the project team to plan and direct the temporary rail station project.

For the remote bridge repair project the sponsor's works department formed an integrated engineering and management team between its staff and a leading firm of consulting engineers established in the region.

The consulting engineer was appointed project manager for the execution of the urgent and unexpected project. The consultants established a site office to manage the contracts and supervise the work. The sponsor provided site services. This team was thus formed by combining the two organizations' staff to bring together the engineering and management expertise required.

For the temporary power line the contractor formed the engineering and management team for the project, with the inclusion of some specialist subcontractors' staff. All on the project were located in the contractor's offices. This team thus also combined engineering expertise and project management.

For the railway reinstatement case the sponsor followed their procedure to appoint an incident manager to coordinate actions on the work and maintain traffic. The project team was formed from members of the sponsor's project department. The head of projects was initially the project manager. The senior engineer who had been project manager of the urgent and unexpected nearby embankment project took over when that project had been completed.

In the 9/11 pile case the team evolved as the leaders of the emergency services personnel on site and the other stakeholders recognized the value of expert coordination of their own and the additional large-scale and specialist resources and supporting services required. The senior city engineer on site with expertise in the use of the key resources, city procedures and a previous emergency repair who had provided the first leadership became recognized as the project manager.

The project teams were thus formed from the staff available and were contingent in different ways on the conditions and nature of the projects, as shown in Table 4.2.

Table 4.2   The twelve cases: Formation of the project teams

Case	Project team	Project manager
New TV business	Sponsor with two partners diverted staff to form a dedicated virtual project team	Sponsor's established head of internal projects together with business manager
Temporary rail station	Sponsor's regional property services manager and top staff diverted full-time to the project	Sponsor's regional head of property services
Thames bank raising	Skeleton staff from long-term project and teams of consulting engineers' staff for each section of the project	Deputy leader of the sponsor's long-term project team
Flood diversion	Senior staff from sponsors' long-term project teams and consulting engineers' team	Senior engineer from sponsor's long-term project team
Ouse bank raising	Sponsor switched consultant/ contractor team from long-term improvement programme	Sponsor's project engineer in charge of long-term improvements programme
Viaduct propping	Sponsor's senior staff diverted to the project full-time	Sponsor's operations director
Embankment stabilization	Sponsor's regional projects department formed own small team and switched a contractor from other nearby work	Senior engineer from sponsor's regional projects department
Remote bridge repair	Sponsor and consulting engineers formed a joint engineering and management team	Consulting engineer
Temporary power line	Contractor formed team with consultants and subcontractors	Contractor's director
Aire banks restoration	Sponsor switched consultant/ contractor team from long-term improvement programme	Sponsor's project engineer in charge of long-term improvements programme
Railway reinstatement	Sponsor formed own small team to work with contractor chosen to undertake future maintenance and small works	Initially sponsor's head of regional projects, then senior engineer from completed Heck embankment project
9/11 pile removal	Leaders of the fire, police and others on site became a coordinated team	City's deputy chief of design and construction accepted as leader and de facto project manager

Team Provenance and Unexpectedness

Resources to form the project teams for these urgent and unexpected projects were therefore brought together in three different ways:

- *Diverted resources* – teams formed by the diversion of resources already employed by the sponsors for related planned work, as seen in the flood repair and railway reinstatement. In the classifications suggested in Chapter 3 these cases were classified as unexpected in *scale*.

- *Augmented resources* – teams formed partly as above and partly by temporary employees, as in the remote bridge repair, flood diversion and flood protection cases. In the classifications these projects were classified as unexpected in *timing*.

- *Bespoke resources* – teams formed entirely for the urgent unexpected project, as in the temporary power line, viaduct propping, embankment stabilization, the TV launch, temporary rail station and the 9/11 pile removal cases. In the classifications these projects were classified as unexpected in *nature* or in *probability*.

Table 4.3   Classifications of the unexpected

	PROJECTS CLASSIFIED AS		
RESOURCING	**Unexpected in *scale***	**Unexpected in *timing***	**Unexpected in *nature* or unexpected in *probability***
Diverted resources	Aire bank restoration Railway reinstatement Ouse bank raising		
Augmented resources		Remote bridge repair Flood diversion Thames bank raising	
Bespoke resources			Temporary power line Viaduct propping Embankment stabilization New TV business Temporary rail station 9/11 pile removal

The number of cases available is too small to establish any statistical pattern, but a simple qualitative observation from comparing these cases is that for the urgent projects caused by events unexpected in *timing* or *scale* (Table 4.3) the project teams were formed by drawing on expertise and supporting resources already employed for similar work planned by their sponsor. Dedicated project teams and managers could be established at the start of those projects. In the other cases the urgent work caused by events unexpected in *nature* or *probability* called for resources not already employed for the type of work required. Formation of dedicated teams followed recognition of the need to empower project management expertise. Forming a project team to start to respond to an unknown unknown was different to forming a team to start to respond to a known unknown.

Chapter 5
Project Implementation

Stakeholders

The stakeholders and sponsors remained unchanged through all the twelve cases, including those longest in duration. A change of management of the work was threatened in the 9/11 pile removal case as the range of interested parties rapidly extended to the state and federal governments, but on advice from their engineering advisers the federal government pushed back from taking over the work and abandoned their wish to send in other national contractors. The city authorities thus continued in the role of sponsor through to clearance of the site.

Objectives

The objectives agreed initially were followed through in all the cases, with the only exception being a change in the 9/11 pile removal case from initially trying to rescue casualties to sifting and searching for identifiable remains. In the flood damage and flood protection cases some details of the work were adapted to suit local stakeholders, but the objectives remained constant. In those cases some details helpful to planned longer-term improvements were included where time allowed, and choices were avoided which could have hampered the longer-term work.

Management

All the asset restoration and preventative projects were undertaken by the departments within the sponsoring organizations responsible for operation and maintenance of the repaired, restored or improved asset. In the rail reinstatement case an incident manager was formally appointed under an established procedure to link the project team to the asset management and operating organizations and so coordinate the restoration work and the temporary provision of substitute services. The reports listed in Appendix 2 of city subway restoration and a large oil platform repair show that appointing an incident manager was similarly established in other organizations operating large assets. The one exception was again the 9/11 pile work where the responsibility evolved with the project.

The reports of the cases summarize how decisions on scope, the employment of resources, timing and risks were made at what became regular meetings of sponsors' management, the project teams, technical experts and advisers. Urgency dictated that many initial decisions were oral and were final. Notable in the 9/11 pile removal case was the reliance placed on making decisions at the large coordinating meetings without recording these in minutes. One observer commented that as actions followed immediately from these meetings, this reliance on oral commitment demanded a clarity in agreeing the conclusions of discussions that is not always commonplace for such meetings with normal projects.

The involvement of top management and stakeholders' representatives was sustained to agree the next immediate actions and accept costs. The reports listed in Appendix 2 of other urgent and unexpected projects not surprisingly also describe frequent meetings as the forum for making instant decisions, for example the city subway repair project and the city power restoration project after cyclone destruction. In some of the cases these meetings were as frequent as twice daily, for instance over the short duration of the railway reinstatement project. On the 9/11 pile removal site the leading representatives of stakeholders came together twice daily initially to argue for priorities and resources. From this emerged acceptance of the need for coordination and then project management.

In all but the two smaller cases, the tasks of managing the stakeholders concerned and managing the resources employed became two complementary roles in order to cope with the intensity of communications needed for immediate decisions. In the new TV business case the chief executive and the project manager operated in this way within the sponsor's organization, as did the steering committee chairman

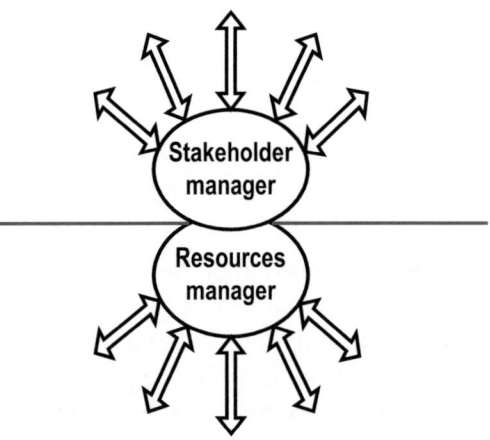

5.1   Complementary leadership roles

and the project manager in the Thames bank raising case. In the remote bridge case the sponsor and the project manager naturally undertook these two roles. Similarly, between the sponsor and the contractor-consultant teams in the flood repair cases and in the rail reinstatement case. The same evolved between the department commissioner and his deputy as the need for the coordination and then management of the 9/11 pile removal work was recognized. In the flood diversion case, the two roles were divided between the local government authority's project manager and the flood authority's project manager. In these various, quite different, conditions the two leadership roles evolved such that all other individuals and organizations dealt with either the stakeholder manager or with the resources manager, but not with both. The two managers worked closely together in complementary roles. They minimized duplication in communications with other parties in the way implied by the red line in Figure 5.1.

The titles and basis of the two roles varied with the resources available and organizational practice. In the rail reinstatement case a senior manager was designated 'Incident Manager' to coordinate the stakeholder's actions, following an established procedure corresponding to practice in UK public emergency services. Use of the same concept and title is reported from the New York subway restoration and the Thunder Horse riser replacement listed in Appendix 2. In Figure 5.1, the general title 'Stakeholder Manager' is used to denote the nature of the role of the person agreeing with the stakeholders what shall be done, and 'Resources Manager' the role of the person managing how to do it.

A separate launch manager was appointed at the start of the new TV business project to plan the handover of the new asset to a new operating organization. In the other cases the completed work was the responsibility of departments existing in the sponsors' organizations.

Resources

Resources and the logistics of deploying them dominated the decisions for achieving urgency.

In the new TV business case each of the three organizations diverted some of their established resources to the development of this urgent and unexpected temporary project, complementing each others' expertise in a joint venture arrangement unusual to the partners.

For the temporary rail station construction the sponsor employed one of their two pre-qualified contractors for supplying and erecting scaffolding, plus call-off local smaller work contractors, all following their normal practice for such work.

In the flood prevention and diversion cases the sponsors augmented their established resources by employing consulting engineers, following their normal practice for projects. They subdivided the construction work to suit the capacity available from contractors already approved for that class of work. For the viaduct propping the sponsor made immediate call on standby contractors to use their equipment on an unusually large scale.

In the remote bridge repair case, no local resources were available. The work required heavy engineering and construction work. The sponsor did not have their own resources and appointed contractors as was their usual practice for their normal projects, except in this case it was by negotiation rather than competitive tender. The work was divided into packages suitable for some regional contractors' immediate capacity and using their knowledge of available material to decide the quickest form of repair.

In the river bank repair and improvement cases engineering and contractor teams and their supply chains already in place for longer-term programmes provided an immediate resource to commence assessing the needs and then to form small teams for the urgent work.

In the temporary power line case the sponsor normally appointed contractors for the packages of work required. For speed in this case an experienced contractor employing consultants and subcontractors undertook the whole urgent and unexpected temporary project, while the sponsor concentrated on coordinating arrangements for immediate substitute services.

In the railway reinstatement case the contractor selected to take over future maintenance work, improvements and small projects was instructed to proceed with this urgent and unexpected project. For the embankment stabilization, use was made of a suitable contractor who had just completed work nearby that was unrelated but had used the appropriate resources.

For the massive heavy wreckage and earth sifting and removal work at the 9/11 pile, four contractors known from previous work for the city were brought in after the start to manage zones of the pile, using local subcontractors and with advice from many consultants.

The resources of partners, consultants and contractors were thus used in various ways in the twelve cases, as summarized in Table 5.1. Common to all the cases was packaging the work to suit the capacity of the available organizations able to respond with expertise in the work required.

Table 5.1   The twelve cases: Key resources and contract structure

Case	Key resources	Contract structure
New TV business	Technical resources of three organizations and sponsor's project leadership	Joint venture between sponsor and two complementary partners
Temporary rail station	Scaffolding contractor familiar with railside work plus local works contractors	Established call-off contracts
Thames bank raising	Project team leaders from longer-term project team and consultant's staff	Work divided into sections to suit already approved small works contractors
Flood diversion	Project team drawn from longer-term project	Three separate contracts each appropriate in size for the resources available from local contractors
Ouse bank raising	Engineering expertise and construction resources drawn from long-term capital works programme	Established contractor and engineering consultants team provided an immediate resource to commence assessing the needs and then form a small team for the emergency work
Viaduct propping	Civil engineering contractor and props supplier already on call	Established call-off contracts
Embankment stabilization	Piling contractor diverted from nearby project	Addition to the scope of the contract for nearby work
Remote bridge repair	Major components adapted from completed temporary materials for another project	Three distinct packages to employ three contractors selected for their track records in the type of work required
Temporary power line	Contractor experienced in inexperienced utility's type of work. Use of rail tunnel route for cables	Contractor with consultants and subcontractors initiated and ran the project
Aire banks restoration	Engineering expertise and construction resources drawn from long-term capital works programme	Established contractor and engineering consultants team provided an immediate resource to commence assessing the needs and then form a small team for the emergency work
Railway reinstatement	National priority for diverting materials and specialized track-laying equipment	Maintenance contractor designated with experience of previous major emergency work available to manage the work
9/11 pile removal	Contractors' expertise in using heavy engineering construction and earth-moving equipment	Four known major contractors familiar with heavy construction

Understanding how to use available resources for unexpected work can be a problem, particularly in intensive work such as at the 9/11 pile. In the animal epidemic case included in Appendix 2, the public authority organization found it difficult to realize that others' local resources, prevented from being used for their normal work by the event itself, could be of immediate use for the urgent and unexpected work required. Excessive resources can be a problem. In the 9/11 pile case many individuals brought resources to the work spontaneously, in response to the news of this massive destruction. This required an additional task of getting the many resources that were surplus to requirements to depart to cease hampering essential traffic.

In all the cases where contractors and consultants were employed their managers were integrated into the project teams to take part in the decision-making process on what tasks were to be undertaken and how to overcome the associated problems, working closely to the cooperative mode known as 'alliancing' as described by Bower (2003). In the rail reinstatement case, collaboration between the sponsor and contractor was facilitated by shared recent experience of working together on a previous major rail reinstatement project.

Reported as being of particular value was the involvement of individual managers who understood exactly how their own organization did and could operate. This allowed urgent and unexpected demands to be met; for instance, to obtain resources rapidly, select a management team, gain approvals and operate temporary cost recording and payment systems.

The cases varied on the impact of the urgent and unexpected projects on existing planned work. In the flood damage repair and flood prevention cases, staff were switched to these urgent and unexpected projects from longer-term related planned work. In the remote bridge repair, temporary power line, new TV business and rail reinstatement cases the normal work of individuals was temporarily covered by colleagues. In the 9/11 pile removal case all such other work was suspended.

Not always recognized at the start of some cases was the need for deputies for managers to match shift-working requirements, especially where the project was expected to be completed within a few days.

Normal procedures of safety plans, independent monitoring and reporting were followed in the cases requiring construction work. At the 9/11 site the parties agreed that the safety standards in the pile sifting should be those applicable to emergencies and rescue rather than construction.

Contract Terms

Cost-based terms of payment or flexible use of variation provisions were used in most of the cases, with some additions of incentives for speed of completion.

In all the cases employing contractors for construction work, the terms of contract chosen were those already approved by the sponsors, for instance in the Thames bank case to avoid delays in asking for waivers of rules or procedures. In the remote bridge repair case, payment was based on a schedule of rates under terms of contract familiar in Australian and British practice, together with incentives for early completion. The bills of quantities basis of payment as operated for the flood diversion work and the Thames bank raising were similar in effect.

For the river bank repair and improvement cases the contractor and consultant teams already employed for longer-term programmes were switched to the urgent and unexpected work under cost-plus payment terms. For the rail reinstatement work the work was similarly ordered under the cost-reimbursable terms of the impending maintenance contract. For the embankment stabilization the work was added under the cost-reimbursable terms of a contract already running for nearby work.

In the 9/11 pile removal case the contracts were made orally, for later confirmation. Terms adapted from previous contracts for emergency repair of a city stadium were used. Employing contractors without competitive bidding was allowed under provisions for emergencies. Payment was based on time and materials reimbursement plus fixed fees, adapted from terms used for previous emergency work.

In all cases, terms of contract familiar to sponsors were therefore preferred, with only secondary adaption to urgency.

Costs

Though the cost of working as fast as possible was not a factor in deciding to initiate these projects, cost forecasts and records were required by the sponsors as the work proceeded in order to plan progress payments and for reporting to stakeholders.

In the London flood protection case the authority to proceed from the sponsor's organization depended on formal approvals of the proposed work, its effects on amenities, possible cost and employment of contractors, requiring a recognized sequence of submissions to committees. To save time the submissions were sent to each committee simultaneously rather than requesting waivers to submitting them.

Waiving rules to get work started rapidly has been reported from other urgent and emergency cases, for instance in California to place contracts without competitive bidding for the Northridge earthquake recovery work. In none of the reported cases was there the political reluctance to waive rules and procedures which has been blamed for contributing to delay in immediate remedial work following other calamities (Philips 2005).

Project Performance

All the projects met their sponsors' objectives. As observed in Chapter 2, urgency was sustained through the lengthier projects such as the Thames bank raising.

Several people interviewed for the case studies stated that the project teams operated very effectively on these urgent and unexpected projects. Individuals commented that the work for their project was not only successful but also exceptionally stimulating and satisfying. Motivation was sustained through several months and even with personnel changeovers experienced in the lengthier cases of the Thames bank raising and the 9/11 pile sifting and removal work.

Contractors previously shown trust when employed on normal projects were reported as responding particularly well to these unexpected demands, for instance in making their best staff available and supporting them in sustained all-hours working.

In some cases communications and procedures in the sponsor organizations changed to respond to urgency, for instance in achieving rapid vertical teams, learning a discipline of accuracy for confidence in oral instructions, rapid selection of resources based upon known performance and simplification of procedures. It was also said that organizational behaviour under the pressure of urgency was closer to the rational principles for managing their companies that advisers had previously said should have applied than their current practice at the time. Surprise was an opportunity for ideas. Urgency precipitated simplicity?

Decisions Critical for Success

Table 5.2 lists the managerial and engineering decisions recorded as critical to success in the twelve cases.

Table 5.2   The twelve cases: Managerial and engineering decisions

Case	Decisions critical to success
New TV business	• Taking on the transmission company and competing commercial company as partners • Appointment of full-time project managers for the project and its launch • Appointment of a project manager able to coordinate staff in different companies and with different cultures • Agreement of all stakeholders to an integrated project plan • Injection of early control of the project, across all parties
Temporary rail station	• Network Rail's Property Groups were used to emergencies and other reactive work and had contractors in place for immediate calls • Temporary footbridges are often required during track work and so a basis for design and pre-qualified contractors were available • Cooperation between all parties. Communications on the project were good as many individuals in Northern Rail, Network Rail, the contractors and subcontractors were used to working together
Thames bank raising	• Project manager appointed under accepted system for inter-departmental projects • Prior briefing of property occupiers and owners • Personal contacting of occupiers and owners wherever possible • Area managers authorized to take the lead with occupiers and owners • Once-only disturbance of occupiers • Division of the construction work into sections suitable for the contractors
Flood diversion scheme	• Agreement to proceed with the emergency scheme when need was uncertain • Quick decisions on the line for the emergency project based upon knowledge of the ground built up for the permanent scheme • Early establishment of project management • Appointment of the project teams already familiar with all parties and the type of project parameters and demands • Employment of consultants already experienced in the area • Employment of known contractors with the resources required • Use of appropriate standard conditions of contract • Agreement on cost-plus terms of payment • Detailed direction and control by the project management team • Round-the-clock working by project team, consultants and contractors
Ouse bank raising	• Strong cooperation between the sponsor, local authority, police, fire and rescue services during and immediately following the flood event • Professional competence of all staff • Use of the contract provisions pre-arranged for planned works • Use of a national procurement contract to obtain a large quantity of sheet piles in a short timescale • Competence and willingness of consultants and contractors

Table 5.2 **The twelve cases: Managerial and engineering decisions** (*continued*)

Case	Decisions critical to success
Viaduct propping	• Highways Agency, Connect Plus Services (CPS), police and emergency services had well-established links for daily traffic management and emergency planning • CPS quick response for analysis of the potential damage • The availability of contractors under framework contracts • Priority given to achieving partial reopening of northbound M1 lanes as access for propping under the viaduct was possible only from one direction. Priority was given to restoring two northbound lanes for football supporter traffic expected to be passing on Saturday at about 8pm. As a consequence of this priority the speed of the subsequent work of propping the southbound lanes was limited because access was possible only between the completed northbound lane props • Rapid response of propping contractor and availability of their resources • Reopening time not promised too soon
Embankment stabilization	• The project team and contractor were brought together within 18 hours of the first call for their services • To establish good relationships rapidly it was agreed with the contractor and the consultants that they would bring known senior staff to the project • A contractor who had satisfactorily just completed a rail bridge reconstruction project nearby was appointed to undertake site investigation, design and construction to stabilize the embankment and other affected structures • Some rock-fill to start to consolidate the berms each side of the embankment could be started before geotechnical investigation, analysis and design had been completed • For site investigation and design the contractor employed consulting engineers who had considerable experience of similar railway work • The contractor began work on the embankment under a variation order to the existing contract for a nearby site. Following this start, a contract for the Heck work was made under a 'rapid response' version of model conditions for reimbursable payment • Separate consultant also with extensive railway expertise was appointed to check design, supervise construction and monitor safety on behalf of the sponsor • Delivery of the very large volume of rock and topsoil was required through village limited to 14 hours per day, despite the urgency
Remote bridge repair	• Immediate authorization to start the project • To proceed with a temporary structure spanning the collapsed pier using long girders on hire • Piles for a new pier could be driven from the temporary bridge, piers strengthened and the collapsed spans replaced • Ordering materials in advance of appointing contractors • Division of the work between contractors having the necessary capacity • Department's inspection, cost monitoring and contract management staff located on site • Early establishment of communication systems for the project and for all stakeholders and media

Table 5.2   The twelve cases: Managerial and engineering decisions *(continued)*

Case	Decisions critical to success
Temporary power line	• Commitment of utility and contractor staff ahead of a letter of intent • Acceleration of first programme for the work • Completion of conceptual design and agreement within three days • Planning to anticipate problems and remove constraints • Immediate actions to procure critical items, some at the expense of other projects • Use of available materials • Contractor's proposal to string an open power line in a tunnel and alongside an operating railway line • Cooperation of railway and city authorities • Commencing construction before detailed design completed • Use of statutory emergency powers for access to land • Commissioning team appointed early in the project
Aire banks restoration	• Strong cooperation between the sponsor, local authority, police, fire and rescue services during and immediately following the flood event • Professional competence of all staff • Division of the whole flooding problem into subprojects for completion by teams • Use of the contract provisions pre-arranged for planned works • Competence and willingness of consultants and contractors • Regular written briefings to the villagers on proposals and progress, and a daily multi-agency 'help caravan' in the village for residents to get advice and help build personal relationships
Railway reinstatement	• Using a contractor where there is an established relationship and a record of cooperative working • Ensuring decision-making is delegated to team managing the incident • Establishing a realistic reinstatement programme at the outset and resisting the inevitable pressure to have the incident site back into operation earlier than can be guaranteed. It is better to plan realistically and with a fair wind deliver early than the other way around • Clarity of responsibilities at the outset • Lessons of previous project known to sponsor and contractor • Good cooperation with the emergency services and regulator • Both parties immediately appointed a strong team • Sponsor's project manager kept communication lines simple so that the contractor had few other parties to deal with • Appointment of a contractor with all the responsibility and the necessary skills to complete the full work scope • Team-building between all main parties • A realistic programme at the outset to take the pressure off staff who were highly stressed • Employment of one contractor facilitated integrated planning so that some over-head line and signalling and telecommunications work could start during track work

Table 5.2   The twelve cases: Managerial and engineering decisions
(*concluded*)

Case	Decision critical to success
9/11 pile removal	• The Mayor and advisers supported the lead taken by their Department of Design and Construction • The department first estimated that the project would take a year and cost $1 billion. This was a shock to the City authorities, but when accepted provided the cover for the quick calls for resources needed to meet the emerging demands of the search and rescue work • Employing known capable contractors • Agreement on reimbursable contract terms of payment • Close relationships with the media • Exceptional safety procedures agreed for certain conditions, temporarily, with employees and all parties including enforcement agencies

Many of these decisions stated to have been critical to success were recorded in terms specific to a project. From them emerge these recurrent themes:

- Early recognition of managing the urgent work as a project.

- Dedication of full-time leadership and coordination.

- Attention to all potential stakeholders, including the media.

- Employing known, capable resources.

- Team-building of teams not located together.

- Location of project leadership at the work.

- Work packaging to suit resources able to start immediately.

- Adapting procedures to urgency.

- Preparation for handing over to asset users.

Convergence on Normal Practice

The twelve cases exhibited a variety of stakeholder structures, project team organization and deployment of resources. This variety was contingent on the sponsors' roles and resources and on the conditions of urgency and unexpectedness of the work. These projects were started differently from the norm in their industries. Once started, the sponsors mixed innovative and established procedures to achieve speed in getting resources and managing their use.

Note

The analysis of stakeholder relationships in this chapter is based upon Wearne, S.H., Management of urgent emergency engineering projects, *Proceedings of the Institution of Civil Engineers – Municipal Engineer*, 2002, 151(4), 255–63; reprinted in *IEEE Engineering Management Review*, 2005, 33(3), 21–31.

PART II
Analysis

Chapter 6
What's Different?

Project Starts

A project normally starts with investigations of the possible demand for its planned product or service, its money-earning potential and other potential benefits, and also estimates of its possible costs, uncertainties and risks. Various names are used in businesses and public services for this investigation stage, for instance concept stage, feasibility study, pre-project, front end, opportunity identification or appraisal. Whatever they are called, the common purpose of the investigations is to provide the basis for deciding whether to continue to use resources to develop the project any further.

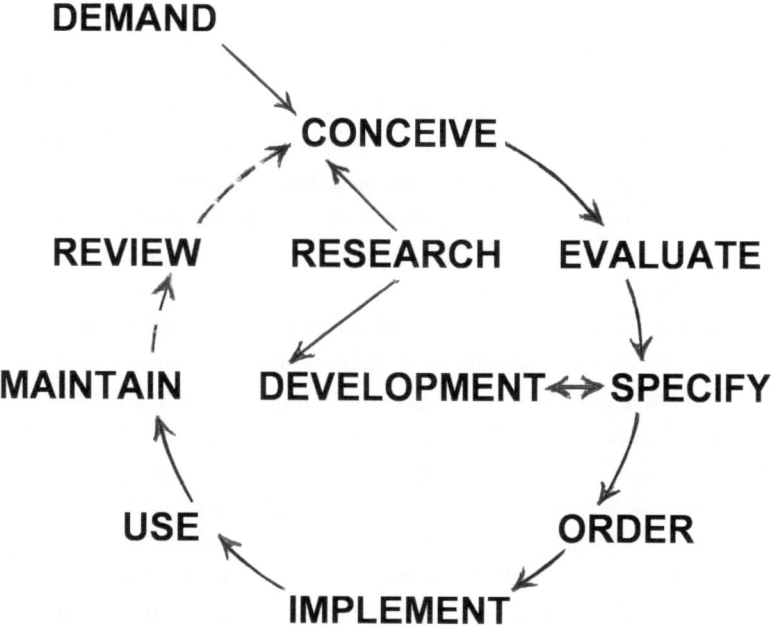

6.1   Project life cycle

How these first investigations should be planned and managed is discussed in textbooks, company guides and research papers. The UK government-supported PRINCE2 handbook (Bentley 2009) specifies a systematic process in which the pre-project stage provides the business case for a project. Cooper (1993), from studies of manufacturing projects, emphasized the value of first giving time to ideas generation, preliminary assessment and concept definition before starting the development of a possible new product. Ideally the concept stage of every new project should draw on the lessons to be learned from reviews of previous projects, as illustrated in Figure 6.1 in a Manchester model of a project life cycle.

From looking at cases of new product development, Nobelius and Trygg (2002) questioned whether there could be one best process for the start of a project. They concluded that the nature and scale of activities such as generating ideas and developing concepts should vary according to the type of project, differing for instance between an incremental development project and an open-ended research project. They and others have emphasized the value of planning the front end decisions to define the project objectives and implementation strategy before commitment to it. The sponsor of a possible project might have different objectives to the users of the potential product or service. So might groups and individuals in different roles and levels in their organizations. And so might their advisers and suppliers. Some individuals or groups under the stress of urgency might want to analyse and plan everything before agreeing any action (PMI 2005). Others to jump into action. Though the data is often ambiguous and not final, from reviewing lessons from a variety of projects Williams and Knut (2010) emphasize that in this 'turbulence' of objectives and interests this investigation stage should be directed to defining what is the *right project* for its sponsor's strategy as well as deciding *how* to do it.

Wherever an idea or perception of a need for a project has arisen in an organization, its investigation is normally the function of an embryo project team or a department specializing in that stage of potential projects. Their results are submitted up the organization's managerial hierarchy under rules for presenting estimates of expected benefits, costs and risks. Decisions are then given downwards on whether to abandon or whether to proceed any further. Continuing further tends to be a cyclic process of repeated and increasingly more detailed investigations, often including stages of consultations with experts, stakeholders and regulating authorities. Starting from first expectations of the demand for a potential product or service, a proposal for a project develops as it moves up and down the hierarchy.

This often lengthy gestation is typical of 'normal' projects. Ideally the result is the unambiguous definition of project objective, scope, plan, risk assessment and budget accepted by all parties that the textbooks, government reports and company guides state are needed at its start for a project to be a success.

The Cases

The cases summarized in Appendix 1 show that their starts did not follow from a normal sequential and hierarchical process. The different levels of functional expertise and higher management which would normally operate level-by-level in series made the decisions on scope, plans, resourcing and control together, typically through frequent meetings or virtual meetings. These meetings brought together the authority to commit resources and the expertise to manage the work. One result was that these projects had an agreed objective at their start.

Concern that the project should proceed as fast as possible in agreement with other stakeholders was perhaps the primary interest of the top managers. Their direct involvement for that reason enabled immediate acceptance of cost uncertainties and other risks as the scope of work and priorities became known during the work. Frequent contact between all levels thus largely replaced the normal sequential and hierarchical procedures in deciding project scope, plans and resourcing. From this flowed differences in how the projects were managed, particularly in the relationships between the project teams and their sponsoring organizations and other stakeholders.

Stakeholders

In a study of managing relationships in construction projects, Bonke and Winch (2000) observed that some influential stakeholders might be remote from a project, some interested stakeholders might have little influence on the project, and some influential ones have little interest. Their studies of these relationships indicate that to determine who will need what attention the sponsor should first identify all the possible stakeholders and then classify their levels of interest and of influence. Stakeholders who have high power to influence but low interest need to be kept satisfied. The others with a high level of interest should be kept well informed.

An obvious difference in the cases of urgent and unexpected projects compared with normal ones was that agreements between stakeholders on what was needed, who would do it and how it would be done were made rapidly at the start. Many stakeholders were involved more as members of the project teams rather than as separate parties consulted before or informed after decisions. In those cases classified as unexpected in *timing* or *scale*, most or all the stakeholders were aware of the possibility of the event and were already parties to consultative planning for emergencies. Links between them already established for a known unknown can obviously help, as in the flooding cases. In the cases classified as unexpected in *nature* or *probability*, the sponsors recognized the need to bring together the stakeholders in reaction to the triggering event. In effect, all followed the recommendations from Bonke and Winch with successful results. Though the projects were urgent they involved the stakeholders in order to secure their early agreement and support of the actions proposed. And 'stakeholders' included the media, whose aid by broadcasting information was reported in several cases.

Stage Gates

During any project the information used for the front end forecasts may be affected by new ideas, opportunities and problems. To establish a discipline of reinvestigating the possible effects of these on the front end forecasts of potential benefits, costs and risks after starting, Cooper (1993) famously recommended that the expenditure of resources on a project should be authorized in stages. Under this discipline, each stage starts only if it passes a managerial 'gate' where the forecasts are reviewed to decide whether it should continue. Passing a gate only authorizes using the resources needed to progress the project to the next gate.

In the twelve cases, the decision to start the project was, in effect, the only gate. The meetings of the project teams with higher management and other stakeholders that explored, defined and authorized the most immediate actions could have been used as intermediate stage gates if the objectives and potential benefits had changed or the forecasts of cost through to completion had become a restraint.

Project Teams

Temporary project teams were formed at the start of these urgent unexpected projects. Managers with experience of delivering any new product or process would perhaps consider it normal to form a project team, as observed by Turner and Müller (2003) in discussing the uncertainties of projects compared with the routines in operations management. The concept of using a project team was probably natural to the sponsors of the urgent and unexpected cases for responding to events only unexpected in *timing* or *scale*. They had people already employed for related longer-term and maintenance work who provided the basis for teams for the urgent and unexpected work.

Managers more used to hierarchical routines and distance in managing operations or services might less readily call together a complete team of experts, resource managers and other stakeholders to identify the possible problems, ideas, opportunities and risks of an urgent surprise. Forming a project team was probably not natural to sponsors in responding to events unexpected in *nature* or *probability*. In the extreme 9/11 pile case collaboration between the groups on the ground to become a team and accept project leadership only developed with recognition of the need for expert coordination of the scale and nature of the work needed.

The teams in the twelve cases were thus formed in various ways, dependent on the recognition of the need and the resources available. None of the sponsors had prior procedures established for forming such teams unexpectedly. All followed Turner and Müller's rationale that a team is needed 'to undertake a unique, novel and transient endeavour managing the inherent uncertainty and need for integration'.

Engwall and Svensson (2002) studied three instances of the formation of a temporary team *within* a project team to overcome an unexpected problem. Two cases were in telecommunications projects, and one in pharmaceutical research. The term 'Cheetah' team was used by them to indicate that they worked fast for a short time. Those teams had not been expected to be needed and were not planned in advance. In one case their leaders and members were drawn from the parent project team, in one from other projects, and in the other a mixture. They were committed to their temporary 'Cheetah' team full-time. These temporary teams within teams achieved their missions. They were then dissolved. Planning was done 'on the hoof'. Engwall and Svensson commented that they were very effective in pushing a project forward, but seldom could be economically efficient. They were thus similar to the cases of forming a separate team for an urgent unexpected project. As they concluded, their major strength was focusing 'all their attention to solving one single, common issue'.

In the twelve cases of urgent and unexpected projects, project teams already established for longer-term work were similarly the parents of some of the temporary teams. Whether with such a parent or not, the dedicated project teams achieved this same focus. They were dedicated to their projects until completed as in the often-cited example of the 'Skunk Works' case in the US aircraft industry in the 1960s. Warren and Biederman (1997) described how that team was deliberately isolated from an organization's mainstream operations to develop a new technology and application as quickly as possible. The advantage expected from this was that the team was left free of their organization's normal thinking, rules and habits. It was a cross-functional team of all the expertise thought to be relevant for a useful result, as advocated by Nobelius and Trygg (2002) to draw together ideas and resources for the front end of a project. So were the teams in the twelve cases. They embraced wide expertise, so were able to draw together ideas and resources from their start. Urgency and quality of product were more important than costs, and they did not need to share resources with other urgent projects. All these strengths could have contributed to the success of their projects.

Organizational relationships and communications are the greatest category of problems stated by people employed on normal projects (Wearne 2014), at least in 'Western' cultures. Problems of organizational relationships and communications might be expected to be greater when the stakeholders and other organizations come together quickly and immediate decisions are needed urgently and unexpectedly. Under the stress of urgency and unexpectedness, the first instinct of some experts and managers could be to want to keep everybody else out of the way so that they can get on with what they know how to do, at the later cost of conflicts in the use of resources, lack of collaboration or public disputes, as Loosemore (1998) observed of crises in construction. The twelve cases showed otherwise. As Loosemore also noted, it is common experience that people respond to an emergency with commitment and with ideas. As similarly observed by Solnit (2009), events akin to emergencies are likely to be motivating, 'bringing unused energies into action'. People interviewed stated that the project teams operated very effectively on the urgent and unexpected projects. One difference from normal projects that share resources was that minds at all levels were openly concentrated on the one project and its agreed objectives. Team members were stimulated into co-axial thinking rather than the hierarchical pyramid. In the lengthier cases this motivation was sustained through many months.

Team Structure

How project teams are and could be structured is discussed in many textbooks, company procedures and research papers. Much attention is given to discussing the extent to which the expertise and other resources needed for a project should be dedicated to that project or should be husbanded in specialist groups to serve all projects. These specialist groups are often called 'silos', implying criticism that they tend to become self-centred rather than seeing themselves as a service. The people in any specialism tend to evaluate their contributions to a project more highly than do the other contributors. They might work to ensure their own survival, like cousin Dawkins' (1976) *Selfish Gene*. These motives can be valuable. The potential value of concentrating people and supporting resources in silos is argued as the means of developing expertise, drawing together experience, offering careers and providing reserves of resources as input to current and future projects. A limitation is that they are usually involved only for one or two stages of a project, and so might not be able to learn from the final results.

The potential value of dedicating resources to a project is argued as the means of concentrating attention to immediate objectives and achieving coherent communications, motivation, continuity and also political and commercial security. These teams can enjoy the completion of their project and learn from the results, but they tend to think their project is unique and incur risks by not using others' expertise. Each team is only temporary and the lessons to be learned from their projects are thus dispersed with the individuals.

This choice in dedicating resources to a project presents a structural dilemma as to whether the management of resources should be concentrated on input and therefore *cost*, or whether on output and therefore *value*. From simulating relationships in project teams facing task uncertainty, Kim and Burton (2002) concluded that a centralized organization will perform better than decentralized if quality of product is important. Much earlier, Lawrence and Lorsch (1967) had observed from studies in US chemical companies that organizational structures evolve differently depending on their business environment, leading to the 'contingency' theory on the appropriateness of structures. They observed that companies which combined expert silos with effective coordination seemed to be best in adapting to environmental changes. Some UK comparisons of engineering organizations in various industries at that time led to the conclusion that structures should be different for largely-repeat incremental projects in predictable conditions compared with those for projects varying in system design or degree of novelty (Wearne 1970). The significant variable in those comparisons was the novelty of the project, so that silo storage of resources would be more effective for repetitive similar projects, to carry reusable experience from one to the next. And *vice versa*; dedication of resources to a project would be more effective for the teamworking thought best for achieving innovations, corresponding to the conclusions mentioned earlier of Nobelius and Trygg (2002) from looking at cases of new product development.

The underlying problem in these structural choices is uncertainty in trying to measure the effects of differences in the extent of dedicating expertise and other resources to each of several projects. The compromise seen in many organizations is a 'matrix' structure with parallel managers accountable for resources and for projects, as analysed by Kingdon (1973) and Galbraith (1977). A matrix structure can apply the best resources to the most important objectives. Or not, depending upon planning ahead what resources will be required and then paying attention to how well resources are being used. The general conclusion by Shenhar (2001) from reviewing the earlier literature and studies of manufacturing and other companies in the US and Israel was that management style, skills and organization should be adapted to the type of project, famously using the phrase 'One size does not fit all' to emphasize that the answer to the dilemma should be project specific. That could lead to different structures for different projects within an organization. This might be in conflict with clarity in the parent organization on project governance and relationships with common services. It could help achieve the objectives of each project. Reality may need this variety. The dedicated teams in the cases were distinct from their sponsors' normal practices. They were reported as successful without any problems being reported on their inconsistency from the normal structures in their sponsors' organizations.

Project Team-Building

A team is a means of enabling people to achieve more than they could alone. Or it should be. A structure is the framework for a team. It is impersonal. This is deliberate in most businesses and public services organizations, so that individuals can be replaced. The effect may be that people employed impersonally in their roles do not share objectives. Or an interest in doing so. In that state they are a collection of careers, not a team.

From observations of various small groups of people, Tuckman (1965) famously described four stages in their becoming teams: FORMING → STORMING → NORMING → PERFORMING. Tuckman and various authors later added a final CLOSING, REFORMING or TERMINATION stage (Rickards and Clark 2006). As Rickards and Clark point out, Tuckman's stages might imply that achieving common objectives must take time. The twelve urgent and unexpected cases indicate otherwise. Those rapidly formed teams were effective immediately. And their common objectives were sustained. The reports of some of the cases show that attention to relationships had been recognized as a critical task. Only in two was attention given to team-building as the teams were starting their work. In those cases not all the team members were located together. The 'Lean' process of collaborative planning was applied in the motorway viaduct repair project reported in Appendix 2. By extreme contrast, in the 9/11 pile case where different independent groups were crowded into one area the recognition of the need for cooperation and direction to achieve a common goal appears to have become the stimulant to becoming a team. The various groups of people there went through Tuckman's four stages without anyone directing the process, notably evolving through STORMING as members of different services at first concentrated on finding their own survivors over the same ground (Langewiesche 2002).

Deliberate actions for 'team-building' are recommended by many advisory organizations, for instance in the advice of the International Project Management Association (1987) about project start-ups. Team-building should help combine individual strengths, develop shared objectives, encourage ideas from all and mould personal and project objectives. From case studies, Atkins and Gilbert (2003) concluded that it was particularly valuable in matrix structures. Team life might affect this. From data collected from many project teams in Israel, Zwikael and Unger-Aviram (2010) concluded that team development improves the success of projects lasting more than a year, but is not worth it for shorter projects. The twelve cases match this distinction. Attention was given to team-building in two of the lengthier cases.

Team-building was obviously not possible in advance before the teams existed. In the cases where the teams were formed from people already working in their organization, some managers or external stakeholders might have been impatient with giving time to team-building. The projects could have been stimulus enough. In all the cases, the objective of the work was clear and the team had no other task.

Possibly also influential on behaviour as teams in the twelve cases was the close involvement of top managers in the projects, following the conclusion of classic studies in the Hawthorne Works in the USA that motivation of employees can increase at least temporarily when top management are seen to be directly interested in their performance (Handy 1987). In the twelve cases the regular meetings of all levels appears to have avoided the development of the hierarchical 'distance' that Wijngaard et al. (2010) observed can develop in business and public service organizations between higher executives and project managers because of the differences in their roles. Explanation of behaviour is always uncertain, even with observations at the time. The reported result from the twelve cases was sustained concentration on their projects.

Project Management

Kingdon (1973) discusses how accountabilities are split in matrix systems used for the sharing of resources among simultaneous projects. The managers of the silos of resources are usually made responsible for the quality of the resources and their output. The project managers are usually made responsible for delivery time and cost. Such a split of responsibilities can precipitate dilemmas between achieving quality, schedule and cost if not enough competent resources are ready before all the committed projects need them. There are many publications on the role of a project manager in a matrix structure, particularly using the term 'weak' matrix to mean a project manager with little influence over the allocation of resources and 'strong' to mean a project manager leading a team largely self-sufficient in resources. Given less attention in the literature is the question of how conflicts in priorities are then resolved. Sudden conflicts could be expected between unexpected and already committed work. Questions of priorities did arise in some of the cases. In all they were forestalled by direct involvement of higher management. This supports the conclusions of reviews of the results of government-funded projects by the UK Office of Government Commerce (OGC 2004) and National Audit Office (NAO 2002) that success needs clear links between a project and its sponsor's priorities and clear senior 'ownership' of decisions with stakeholders.

Similarities to the twelve cases are to be found in a study by Söderholm (2008) of how unexpected events were dealt with in four projects in Sweden. Two were in heavy engineering manufacturing, one in the development of medical equipment and one a public health organization. The unexpected events affected stakeholders' interests, project plans and project members' job priorities and expectations. Söderholm observed the following practices by project managers to try to keep their projects on track:

- Innovations in management actions and procedures, from the start of a project.

- Extensive meeting schedules and short-term coordination.

- Isolation of the consequences of revisions, as much as possible.

- Negotiation with all resource and stakeholder groups.

Contrary to this style of leadership could be the hierarchical view that what some call the military style of 'command and control' would be appropriate in situations of surprise and urgency. Various strands of experience and of research say differently. From reviewing reports of the behaviour of organizations after disasters, Solnit (2009) observed that, in the shock and often isolation from information, distant administrative minds tend to think they should impose command. They expect that the local systems, if left undirected, would be 'anarchic and tolerant of disorder and crime'. Hashimoto (2000) reported that civil authorities after the Great Hanshin-Awaji (Kobe) earthquake were similarly distrustful of on-the-spot decisions. They were wrong. Objectivity and selflessness dominated local actions.

Fire-fighting and other civil emergency services have used the term 'subsidiarity' to mean that the trained leader at an event decides how to use the resources available then and there. As the Dutch might say, support the boy whose finger is plugging the dyke. Knowledge rather than rank is the basis of successful decisions in military 'special forces' operations akin to teams for urgent, unexpected work (Melkonian and Picq 2010). Higher ranks set the priorities. The middle levels organize the provision of resources guided by whatever information they can get from the leaders at the event. 'Empowerment' is the equivalent term in management literature meaning that the principle is established ahead of an event that an individual or team has the authority to match their knowledge of a situation. In Western culture managers are expected to delegate this authority, keep distant and check primarily on *how* authority is used without 'interfering' in *what* is decided (Handy 1987). In the cases studied here, uncertainty and urgency dictated involvement of higher management with the project teams in the decisions on priorities, resources and how to use them. As Weick (1993) observed from studies of various organizations under threat, face-to-face interaction is the means of understanding the demands of rapidly changing situations.

Project Managers

Many studies analyse the role of project manager – by that or other titles. There is common agreement that the role includes coordination of all the work to be done. Under urgency the immediate task of managing what are once-only decisions in coordinating work in progress could leave no time or energy for thinking ahead, maintaining stakeholders' support, communicating the objectives and preparing all parties for handing over the completed work, as was reported in cases unexpected in *nature*. Almost double-headed leadership evolved in some of the cases to be able to manage these tasks. The concept of 360° managing in all directions in a hierarchy is nothing new. The combination required of skill to agree decisions with stakeholders and expertise to direct resources is familiar in practice and in management selection and training. The UK government's preferred PRINCE2 system for managing projects defines the role of 'Project Executive' representing the sponsor and other stakeholders, and 'Project Manager' responsible for delivery, corresponding to the 'Stakeholder Manager' and 'Operations Manager' roles noted in Chapter 5. In the cases studied here, these roles evolved rather than were planned, supporting the Tavistock Institute's classical socio-technic theory that relationships in organizations in reality tend to adapt so as to make their formal systems effective (Handy 1987).

In the case of railway disruption and reconstruction the coordination of stakeholder interests was based on the sponsor's established system of naming an 'Incident Manager', corresponding to 'Stakeholder Manager'. The same is reported in Appendix 2 from the Thunder Horse offshore platform riser replacement and the New York subway repairs cases. In the latter the person already named for that role if needed was not available, but as the principle was established in that organization another experienced manager was able to exercise that role immediately and successfully. Appointing an individual to coordinate all decisions is an obvious need. In those instances the principle was understood before the event. This is normal in training in emergency services. Defined roles also provide a framework for continuity in the handovers in two- or three-shift working.

In none of the twelve cases were the individuals who managed the projects selected because of training for the unusual nature of urgent, unexpected projects. In reviewing the experience and training that can be required for professional command in emergency services, Flin and Arbuthnot (2002) comment that this preparation of individuals is not to be expected in other public services or in businesses.

In the 9/11 pile case the person who became de facto project manager was the one city official there who had the appropriate professional expertise. He had experience of emergency work which though comparatively very minor could well have contributed to his performance and to his acceptance by the stakeholders. His authority in a mélange of aroused group cultures came from their recognizing his expertise in how resources could be deployed. It provides an example of *expert* power effective among groups normally based upon *bureaucratic* or *position* power.

Project Planning

It's inherent in surprises such as the cases that they are started without prior knowledge of their scope or attention to planning how to deliver them. This limitation on preparation is obviously a difference from normal projects. Project definition might then depend on decisions from what may be unrelated stakeholders, and few or none of them experienced or prepared with a plan for more than asset maintenance or interruptions to business continuity.

Notable in the twelve cases was the time taken at the start to plan. Less haste meant more speed.

In the cases of restoring assets and services after damage, the urgent and unexpected work followed the emergency stage of rescue and damage limitation, forensics, press and political interest. The response to the damaging and dangerous emergency was controlled by the professionals, the police, the emergency services and perhaps the military. Compared with the front end study for a normal project, that preceding emergency stage might not allow access to prepare for the unexpected and urgent work to follow. Only at the handover from the emergency can the sponsor investigate much of what needs to be done to restore or replace the asset.

During what we call normal projects the relative importance of performance, time and cost tends to change. A simple instance is that after agreeing a contract for the work, interests in the bargain tend to diverge. What the work is costing becomes increasingly important to the supplier; quality and delivery increasingly more important to the customer. The opposite divergence might be expected if a project was initially agreed to be urgent. That first agreement to urgency might decay at alarm over its possible cost to all parties, or new risks to performance. This did not occur in the twelve cases. Urgency persisted.

Many normal projects are started with stress that their final cost should not exceed a stated maximum. During project implementation the sponsors tend to try to 'buy time', that is pay to accelerate the work to demonstrate progress or to overcome delays, overriding the financial case for starting the project. In the cases studied here, buying time was the criterion for planning the work.

Fast Track(ing)

A guide published by the European Construction Institute (ECI 2002) defines 'Fast Track' as when reduction of time is the principal driving force for one or more stages of a project. The report specifies how to achieve and sustain schedule reduction stage by stage through an industrial project. It specifies that the main techniques which can be applied to speed up delivery are the overlap of project stages, overlap of work packages, integration of client/contractor project teams, early decisions, if necessary with limited information, acceptance of additional cost risks, uneconomic staffing and sustained management support for all the above. How far to apply these techniques should follow from the definition of urgency in the objectives of a project. For instance, Larken (2002) reports that, though urgent, the unexpected demand to recommission a warship had to await the proper completion of refit engineering work.

The ECI report acknowledges that none of these techniques are unique to fast track projects, stating that the difference between a fast track and a normal project is the extent to which they are applied. Their recommendations are in terms of a project of defined scope. The twelve cases show the same applies to unexpected and unplanned projects.

Resources

In best practice the resources a project will need should be allocated and prepared according to an agreed implementation plan and budget. None of the resources available in the cases were prepared for the demands of this urgent and unexpected work. The resources that were used were those available which could be deployed within the practical limits of space and technically possible speeds of work. Authority to employ and pay contractors for services or goods was needed before the scope of work and its likely costs could be well defined. Plans and cost estimates developed during the work as the needs became apparent.

Urgency displaced competitive tendering as the normal procedure for purchasing goods or services. Normally the ideal basis for asking potential contractors to compete for work is a full, final and unambiguous specification of what is to be delivered, and risks affecting costs are in the contractor's control (Bower 2003). Without this basis the prospective contractors could not be required to compete on equal terms and the best of them at managing the type of work offer the lowest price for delivering what the user needs. In all the cases where contractors were employed they were already known or assessed as capable and willing. Cost-based rather than price-based terms of payment were used, as is recommended by Ward and Chapman (1994) when the scope or conditions of work are uncertain. These terms permit flexible use of variations appropriate for the 'onion swell' of growth of the amount of work if found to be required as it progresses (Carter 1991). In effect, the customer directs the contractor's use of resources.

This cooperative working with contractors, use of cost-based terms of payment and flexible use of variations used in most of the contracts corresponds closely with the 'alliancing' concept of employing contractors as partners so that all parties concentrate on the needs of the project and jointly anticipate and manage the risks rather than dispute their consequences (Bower 2003; Zhang and Flynn 2003). The 'alliancing' mode of integrated management between customer and contractor is then appropriate, as concluded by Le Masurier et al. (2006) in studying terms of contract used for reconstruction following an earthquake.

Using known terms of contract in the cases for the rapid employment of key resources meant that all parties were familiar with them. This also supports the observation from the cases that when urgent it might be quicker to follow established procedures rather than take time trying to get agreement to waive them. By contrast, to motivate contractors to work fast, 'A + B' incentive terms of contract were reported as used by the California Transport Authority for the extensive replacement of collapsed major freeways after the 1994 North Ridge earthquake (Baxter 1994). Under those terms, contractors bid 'A' amounts for the expected quantities of work and a 'B' number of days for the work. Similar terms were reported as used by the Oklahoma

Department of Transportation to motivate contractors to work fast on the Webbers Falls bridge reconstruction project (Bai et al. 2006). In those cases the time needed for clearance of wreckage and design of replacement structures gave time to introduce these incentive terms of contract.

Changes

The lengthy gestation of many normal projects makes it likely that the initial information used to investigate value and cost will change as the project goes ahead. The consequences of coping with changes in requirements or scope during the work for a project was the absolute majority response obtained by Mantel (2003) in asking US project managers what was the most important single problem they faced in their day-to-day work. Changes are not necessarily negative in impact or 'tragic', as pointed out by Terwiesch and Loch (1999) in a study of change orders in an engineering organization; 'rather they are part of the process of development of a project and coping with uncertainty'. New information and problems affecting the scope, speed of work or cost are all reasons for proposing changes. To manage them textbooks and company rules state procedures for assessing and accepting or refusing changes to the scope or schedule of work, in principle similar to the stage gate discipline but often thought to be too slow for decisions on detail. Success of these procedures depends upon anticipating a possible change early enough to assess its potential risks to achieving a project's priorities between performance, time and cost.

An underlying problem can be that changes can be too numerous during the implementation of normal projects because the first decisions on objectives, scope and risks are not given the best of attention because of at least the subconscious assumption that those decisions will be revisited. The experience that changes will occur later might also deter initial concentration and commitment to decisions, eventually causing corrective scope changes, repetition and duplication of work, and so delay and greater cost. Many delays and other losses due to changes are thus due to poor practices and decisions made prior to the selection of a project (Wearne 2014). To counter this, imposing a policy of 'No Change' is sometimes thought ideal in project implementation to try to enforce the completion of the work on time and within budget. In practice it means instituting procedures to deter changes.

Managing changes of scope was not reported as a problem in the twelve cases, probably because they started from only a provisional or limited definition of the work expected. The objectives agreed initially with the stakeholders did not change. The scope of the work evolved through each project, so they were akin to research rather than normal projects in their industries. In the 9/11 pile case, for instance, from initially expecting to rescue casualties to fine searching for remains and then making the site clear and safe. In the flood and rail damage cases the objectives were constant, but the detail of the work was adapted to suit local stakeholders. Detail helpful to longer-term work was included where time allowed and choices avoided which could have hindered the longer-term work. Scope evolved during the twelve cases because time had not allowed the normal front end studies. Following Winch's definition, project management was thus a process of reducing uncertainty by managing the evolution of scope rather than managing changes (Winch 2002).

In none of the twelve cases were alternative solutions to an expected problem initiated in parallel in case of need. In such a project, integrated configuration management would be needed over the alternative solutions so that all were directed towards a decision on which to select at a project gate along the critical path of time.

In the Darwin case following extensive cyclone damage, power system resources were drawn in from neighbouring Australian states. Each state's team undertook a zone of work. Several brought different standards of engineering and hardware. Requiring conformity could have delayed their urgent work. The results could therefore be only temporary, to be changed later to have consistent assets.

What's Not Different?

Urgent unexpected projects have to be rare in business and government to be tolerable. They are 'outliers', outside the normal range. In his original observations on people he called 'Outliers' as they achieve so much more than others, Gladwell (2008) identified that they were distinct in one significant factor. In the cases, the distinction was the combination of urgency and unexpectedness. Uniqueness was partial. Otherwise the cases support many of the established lessons of managing normal projects, for instance the value of an experienced leader, establishing agreed objectives and simple priorities, employing a qualified and prepared team, open communications and involvement of all parties in early realistic scheduling, using appropriate contracts, attention to safety, positive relations with the media, controlling where the risks arise, regulating changes, keeping records and being mindful about the process as well as the product. Eastham (2002) observed this in commenting about the earlier UK cases. Eijkenaar (1997) commented similarly about the management of recovery work after disasters.

Most of the technical work required in the twelve cases was also similar to other projects in their industries. The differences in managing these urgent and unexpected projects were in the concentration of authority and leadership dedicated to the project linking the sponsors, other stakeholders and project teams; the simultaneous involvement of all levels of management in once-only decisions; reliance on oral commitments; making maximum use of all usable resources; and the immediate acceptance of cost uncertainty. These differences are particular to the combination of unexpectedness and urgency. If a project is unexpected but not urgent, it could be defined and budgeted in the normal way before starting it. If it is already expected and defined but becomes urgent, its schedule can be accelerated as exemplified in 'fast track' practice. If a project is both unexpected and urgent it might seem obvious that commitments may have to be made orally, all possible resources used and the consequent cost accepted. Less obvious until reported from the case studies was that all interests should be represented at the start in an integrated management structure so as to provide a dedicated system of project stakeholders working with dedicated project leadership.

Chapter 7
Lessons

Lessons for Who?

The comment was made at the start of this book that the chance that any one person not in the emergency services will ever manage urgent, unexpected work is small, and it's not possible to know who should be prepared for this. Their organizations can be prepared. Business and government organizations can plan how they will authorize and support future urgent, unexpected work when suddenly needed to take advantage of surprises ranging from new business opportunities to the protection or restoration of assets.

The lessons that may be of value to them listed here are drawn from the previous chapter and the reports listed in Appendices 1 and 2. Each lesson might appear quite obvious to readers with related experience. They might not be so easy to apply if contrary to practice and habits. And not all are relevant to any one occasion. These lessons are therefore not rules. They are a guide, a check list.

To be brief, each lesson is stated as a heading. Comments follow as bullet points for readers with the time for them.

Policy

AGREE WHAT IS 'URGENT'

- If urgent means faster than normal, how fast? As speed generally increases expense, at what cost to whom?

- Urgency stated only qualitatively can have different meanings to the people involved. Expressing urgency quantitatively in money terms might avoid the problem that what is considered to be urgent by some might not be seen as urgent by others. And that urgency has to be maintained when other work demands resources.

- In the twelve cases, 'urgent' had the extreme meaning of working as fast as possible within the limits set by logistics and safety. That established a clear objective. If this is not meant, uncertainty should be avoided by stating the money value of the time which justifies the extra cost of working faster than normal. Money figures influence better than logic.

- Where the word 'urgent' is used to mean faster only to buy time at a calculated rate, the term 'fast track' is a better guide than 'urgent' to the project leaders and all concerned.

- Many normal projects are started while some differences in objectives between stakeholders remain unresolved and internal commitments are incomplete. Some normal projects lack the defined start advocated in textbooks and guides on project management. This can be called risky haste, not necessarily due to urgency.

- Distinguish urgency from emergency. The word 'emergency' has a specific meaning in public protection and services. The word is best limited to life-threatening situations where the resources available are insufficient to preserve life.

MANAGE URGENT AND UNEXPECTED WORK AS A PROJECT

- Help inexperienced stakeholders to recognize that urgent and unexpected work requires the project mode of management so as to agree the objective, define roles and responsibilities and manage scope, risks, priorities, resources, communications and reporting through to completion and handover.

- Authorize one person as incident manager or project manager or whatever title is effective with the stakeholders and with agreed responsibility for decisions on project scope and execution.

- Establish among the stakeholders where oral agreements will be honoured. Ascertain who later will accept that greater cost was rightly incurred in working faster.

- Record why the work was agreed as unexpected, for acceptance of liabilities and for future preparedness.

CHECK ACTION ON 'URGENT'

- Check immediately on how leading individuals are preparing to start actions agreed as urgent. Waiting until they should start might be too late an indicator of what they understand by 'urgent'.

ALLOW FOR OPTIMISM BIAS

- Optimism about the potential advantages of any action relative to its possible problems as observed by Lovallo and Kahneman (2003) in a variety of projects might occur in managing surprises. Studies by Flyvbjerg and colleagues (2011) of major projects showed that forecasts of the value and the costs of projects tend to be optimistic because of inexperience, enthusiasm or self-interest. In a study for the UK government on the causes of cost overruns of public projects, the consulting engineers Mott MacDonald called this 'optimism bias'. Predictions of the value or the costs of working faster than normal are just as likely as normal estimating to be affected by optimism bias. If the decision is urgent then everybody concerned needs to recognize these uncertainties and to agree that decisions can be based only on estimates that are available and sufficiently valid.

Stakeholders

IDENTIFY THE STAKEHOLDERS

- Include not only asset users and owners but also, directly or indirectly, their employees, service providers, insurers, local government, emergency services, social services, charities, regulators, the media and the public and others dependent on them and those they depend upon.

INVOLVE THE STAKEHOLDERS

- Included parties tend to become helpers. Excluded parties tend to become hostile. People ignored, disrupted or displaced because of intangible and unexplained risks tend to become unhelpful.

- Establish a stakeholders' forum for representatives to meet regularly.

- Explain the particular demands of urgent and unexpected work, for instance for agreement to immediate decisions that are final. Explain the risks and risk management plan.

ASSESS STAKEHOLDERS' INTERESTS AND RESOURCES

- Stakeholders differ in their interests, risks, perceptions, depth, resources, their potential contribution, their priorities and their commitment to the objectives, subsidiarity and timing of interest. Individuals, local organizations, central organizations and supporting agencies might differ seriously in their objectives, priorities and resources. Not all stakeholders might understand the need for urgent work or work unexpected in *nature* or in *probability*. Stakeholders might underestimate work unexpected in *scale*. Stakeholders might not understand the change of priorities needed for work unexpected in *timing*.

- Relationships with stakeholders might be the most difficult task of making a good start to urgent and unexpected work. Some influential parties might be remote from the problems, some interested parties have little influence on achieving their solution and some influential parties have little interest. To determine who might need what attention, map all stakeholders' levels of interest and their ability to help – see the diagrammatic example in Figure 7.1.

	LEVEL OF INTEREST	
	Low	High
Low	UPSTREAM COMMUNITIES	DOWNSTREAM COMMUNITIES
	CITY HISTORIAN	CITY COMMUNITY
	COMMUTING CITY WORKERS	REGIONAL MEDIA
	CITY TOURIST INDUSTRY	RIVER POLICE
	ENVIRONMENTALISTS	FIRE AND RESCUE SERVICES
	CITY ARCHAEOLOGIST	RIVER PILOTS
POWER TO INFLUENCE		REGIONAL TRANSPORT AUTHORITY
	GENERAL CONTRACTORS	RIVER FRONTAGERS
	REGULATORS & INSPECTORATES	PROPERTY OWNERS
	MILITARY	
	RIVER CONSULTANTS	
	CENTRAL GOVERNMENT	
High	SPECIALIST CONTRACTORS	CITY AUTHORITY

**7.1   City flood walls rebuilding:
Stakeholders' levels of interest compared to power to influence**

Source: Wearne (2002), after Bonke and Winch (2000).

ASSESS STAKEHOLDERS' MANAGERIAL STRENGTHS AND LIMITATIONS

- Identify who in stakeholders' organizations can authorize work and how they will pay for what. Many businesses, civil organizations and individuals might be unprepared for the concentration of their managerial authority needed to make immediate decisions to start urgent work for them that is unexpected in *nature*. Local government and regulators' staff likewise might be overwhelmed by work unexpected in *scale* that has to be coordinated and checked over much of their area.

- Some stakeholders' urgency to capitalize on a new opportunity or to improve or recover infrastructure might be only partly influenced by calculation of the financially best speed of expenditure. It might be influenced by business, social and political objectives not shared by others.

- Stakeholders' prior plans and resources for managing small surprises might not only be overwhelmed by the demands of large-scale work, but might also be inappropriate or unsuitable because of restrictions on access, resources and dependence on others.

- Stakeholders might need to be briefed on how the roles, authority and limitations of the emergency services vary with events.

SUSTAIN STAKEHOLDERS' INVOLVEMENT

- Anticipate that stakeholders' initial interests tend to diverge during any agreed programme of work. Few stakeholders will have experience starting an urgent and unexpected project and that it demands immediate commitment of sponsor, owners and all others. Their agreement to urgency might diffuse under the impact of stakeholders' reactions, emerging cost or other consequences.

- If the meaning of urgency becomes doubtful, institute stage gates for decisions to review project scope, plan, risks and budget and adjust these to any problems or changes in objectives.

Team Formation

BUILD ON PRIMARY KNOWLEDGE

- Bring together all the expertise required around those first on the job. The resources immediately available govern the first actions. Continuity of their unrecorded knowledge can be vital to speed in adding more resources effectively.

FORM AN INTEGRATED PROJECT TEAM

- Expect that the intensity of communications and dependence on oral commitment will require a dedicated full-time core team.

- Embrace stakeholders' representatives, experts and managers of resources as partners. Local knowledge and commitment are valuable.

- Locating all a project team together physically might be commonly limited by the financial and social cost of doing so temporarily. 'Dedication' more often means that individuals and groups are allocated full time to a project but remain in their separate locations. Plan team-building of team members not located together. The twelve cases show that team-building can develop rapidly, as the continuity and concentration achieved through dedication to one project is stimulating and generates ideas and motivation. Urgency can provide a common goal for a team, but it should not be assumed to be enough to inspire constructive relationships between people brought together rapidly and only temporarily.

PACE THE FIRST DEMANDS ON INDIVIDUALS

- Prepare for the long haul. What is urgent might not be brief. People might have to be attached to a project about 70% longer than they first expected.

- Pace at all levels in organizations, from 'stakeholder manager' and 'resource manager' to all supporting them. An initially high response to urgency and other motivation might soon be followed by loss of performance, as the Yerkes–Dodson pattern cited by Flin and Arbuthnot (2002). A team formed for a limited objective might not perform well if stretched by a further task.

- Shift handovers need time to inform, review and replan.

- Only temporarily ask individuals or organizations to undertake work that's out of their depth.

- Don't assume that normal roles can be sustained in abnormal conditions.

WELCOME EXPERT GUIDANCE

- Invite experts to anticipate latent hazards, for instance the removal of the weight of the 9/11 pile ran the hidden risk of ground heave as well as the more obvious hazards of wrecked structures and back flooding from subway tunnels.

Planning

GET AGREEMENT ON PRIORITIES

- Assess the reliability of the agreement to urgency.

- Assess the total situation, in order to define the critical resources needed. Test initial information if time permits. Lack of information about the extent of work required is common to most (if not all) projects that are triggered by surprises, often because communication systems are not in place or are overloaded.

- Assess whether the precipitating event is isolated or might be followed by repeats or knock-on events.

- Ideally, the first stage of any urgent work needs early guidance from a longer-term plan.

- Plan work packaging to achieve speed of completion. If resources are limited, ascertain whether to give priority to completing subsections sequentially or all together later.

- Allow for drawing off of resources to related events.

- Some stakeholders might be ready for work to proceed earlier than others, but care is needed logistically because the boundaries of utilities and others differ. Sharing facilities or work between stakeholders might leave one stakeholder with the risk that the others do not share the cost.

- Check whether agreement on paying for urgent work to repair or replace an asset might vary according to whether the damage was mankind-made, mankind-incited or mankind-innocent.

- Insurers will want to control the authorization of proposed repair work.

- Plan for return to normal users and operators.

ESTABLISH THAT FIRST DECISIONS WILL BE DECISIVE

- Make clear to all parties that urgency demands decisions which are final, instead of the normal cyclic process of assessing ideas, needs and choices, provisional decisions, development, evaluation, review, formal approvals and design freezing.

- Prevent provisional decisions from diluting the agreed objectives.

- Technology and logistics govern choices. Involve all relevant expertise and responsibilities.

- Concentrate on facts before preferences. People tend to see a need to do what they know how to do.

- Be quantitative about scope as far as possible.

- Stage-gate meetings can overcome distant communications between levels of management. This might have occurred in meetings in the twelve cases, but was not reported.

ASSESS WHETHER ALL THE WORK REQUIRED IS URGENT

- Most stakeholders can be expected to want priority for their interests. Not all of these might be urgent.

- Conversely, some urgent work might depend upon other apparently non-urgent work.

- Some stakeholders might hope to get work done that is not due to the precipitating event.

CONTROL INNOVATION

- Investigate whether to provide a quick temporary project and to interim standards with built-in provisions for later elevation to a higher standard.

- An unexpected project can be an opportunity to make changes to an asset or procedures. Consider whether an unexpected project should be an opportunity only for already agreed innovations to assets, products or procedures which had already come to the end of their economic and safe life, or whether it presents opportunities for greater improvements.

- In two of the cases of urgent and unexpected projects to restore severe damage it was agreed not to take time to improve or otherwise change the design, as the asset had been up to current standards and requirements. This is consistent with McDonough and Pearson's (1993) conclusion in their study of product development projects that urgency is more likely to be achieved by relying on familiar technologies rather than by combining innovations and urgency. That conclusion appears to conflict with the observation in one case that the urgent and unexpected project was an opportunity to make a change that promised to save time on the project and future work. In that case the change had already been thought through.

ISOLATE A CRISIS WITHIN THE WORK

- A crisis can occur in executing a part of urgent work if its objectives are no longer achievable, for instance because of lack of resources or limits for safety. If the facts justify it, a crisis might be best managed as a 'Cheetah' subproject within the project. In effect, the solution to a crisis is a replacement project, to be evaluated and then, from the evaluation, to be abandoned or to proceed if justified.

- People on a project might not know or agree that they are approaching a crisis. A person part way through some work for the project might state confidently that it will be completed on time, but others dependent on using the results might feel certain that what they will get will be seriously late or unfit to use. Ideally, the project manager would be in regular and informal contact with everyone on the project and they would give early warnings of anxieties and doubts. In bureaucratic organizations it might be difficult to get agreement that a crisis is approaching, particularly early enough to minimize its effects. The result can be that by the time a crisis is apparent a decision on it is usually urgent and personal relationships might be deteriorating. All this can deter fact finding. Isolating the crisis can motivate its team to resolve it.

- A crisis could have occurred at the 9/11 pile after the initial work had shown that there was no further prospect of rescuing people still alive, but in that case the sifting, searching and removal work continued as urgent because of the community's need to find evidence of the casualties. Without that continuing driver the remaining work of making safe and clearing the pile would probably have been reorganized for completion at least cost.

Resourcing

SELECT RESOURCES

- Establish a single point of resource mobilization and coordination.

- Employ known, capable consultants and contractors. Do not expect best help from those previously treated distantly.

- In each organization, utilize individuals who understand how their organization works.

- Stored kit might come quicker than kit pulled off other work.

- All organizations need to resource for sustained 24/7 working, including management deputies.

- In public emergency work, dedicate a coordinator of help from volunteers and charities. After urban disasters such as Kobe and at the World Trade Center, resources of all sorts clamoured to be useful, as an expression of concern.

PACKAGE THE WORK TO SUIT RESOURCES AVAILABLE TO START IMMEDIATELY

- Provision for emergency work to protect and recover industrial and public service systems and services from damage vary, logically according to the risks. Protection and recovery from minor damage is recognized as part of the regular tasks of maintenance resources. Similarly, allowances for 'contingencies' are made in allocating the resources for new projects. The managers of maintenance work and new projects are expected to prepare for these 'known unknowns'. They can learn from them. They do not usually provide for 'unknown unknowns'.

- Switch suitable resources. Look for resources themselves stalled by the precipitating event.

- Coordinate multiple demands on the same resources. Even in what appeared to be only a local, though severe, emergency, a logistical problem was created by the separate employment of many contractors for rubbish and dirt removal immediately after the 2004 flooding of Boscastle.

- Allocation of resources might depend upon wider and political priorities.

- Ascertain who will meet the costs incurred to others by the loan or sharing of resources.

ADAPT CONTRACT PROCEDURES TO URGENCY

- Consider using familiar terms of employment for consultants, suppliers and contractors. Select terms appropriate to the urgency and the uncertainty of the work.

UNDERSTAND THE BUSINESS, SOCIAL AND POLITICAL PRIORITIES FOR RESOURCES

- Large-scale nationally urgent work might precipitate regional conflicts in the demand for resources, access, space and liability for costs. Consensus in using resources is preferable, but this might be beyond immediate possibility to respond to many human, physical, environmental, social and economic demands. After a catastrophic event the government might take legal powers to direct public and private stakeholders in their use of resources. Decisions might be controversial, differing from what locally might seem most important, viz. whether to aid or to abandon severely damaged or cut-off services.

Management and Direction

MANAGE WHERE THE FACTS ARE UNDERSTOOD

- The more that work is unexpected in *nature*, the more control should be at the action. Ill-informed distant authority cannot know how best to use resources, especially if influenced by political fear or interest in being seen to be taking action. Exhibit success of subsidiarity when the actions needed cannot be forecast. Agree objectives with those on the ground.

- Set up a common exchange of information between all levels. Involvement of top management in starting a project can have the 'Hawthorne' effect that their evident interest and support motivates all parties to maintain the agreed objectives. Concentrated attention of all levels of management together is motivating not only because of direct contact of levels but also because all are involved and are aware of the reasons for decisions.

- Report through the usual channels. Do not bypass a hierarchy.

ESTABLISH AUTHORITATIVE LEADERSHIP

- A twin-headed control team might be required to have the capacity to achieve final decisions first time. Complementary roles of stakeholder management and resource management work if closely linked and all parties communicate with one or the other.

- Establishing priorities, as far as possible in consultation with political authorities, businesses and other stakeholders, should be one role in this structure, plus relationships with the media and the wider public. Directing the urgent operations would be the other role, if possible using existing services, plus planning the handover to owners and users.

SELECT APPROPRIATE PROCEDURES

- Under urgency situations communications need to be final and precise. Clarify and confirm decisions orally at a meeting not afterwards. Making first decisions final can stimulate good decisions.

- Follow established procedures as far as relevant and time permits, for instance brief definitions for roles. Use relevant existing resources and standards for attention to quality and safety. Limited time might be better used following established procedures than asking for waivers from authorities who do not directly share the interest in urgency. Work the procedures already in place, for instance in one case by obtaining in parallel agreement from various bodies which would normally be sought from each in sequence. Adapt supporting procedures for switching from hierarchical to collective decision processes.

- Enable rapid purchasing. In their advance planning for surprises, organizations should have established procedures for rapid commitments, for instance authority to enter into immediate contracts for goods and services.

- Get early agreement to procedures for handing over completed work to owners and users.

FORECAST POSSIBLE COST

- Forecasts of possible cost are needed early to arrange payment for rapid use of resources. Though calculations of cost or financial benefit are not the basis of defining the work as urgent, keep records of who did what, where and when needed for payment and accounting, and for learning lessons.

- The more sudden and short a demand for goods or services the greater are likely to be the costs of discontinuity. Some suitable but idle resources might be available to use for unexpected work, but this might require paying some of their accumulated standing costs.

Learning from Experience

LOG AND LEARN FROM EVERY SURPRISE

- Record events from the start. The initial decisions are the most influential, at the time and for lessons to be learned. Writing notes makes minds think.

- Ideally record choices and reasons. Agree facts. Invite comments.

REVIEW THE LESSONS TO BE LEARNED FOR MANAGERS

- Lessons are not learned until they change behaviours.

- The lessons to be learned from these different cases in different industries are surprisingly similar. All are potentially helpful. The lessons overlap. Which are relevant might be little or nothing to do with the type of industry or the cause of the surprise. Experience reported from one industry or service is potentially relevant to any other. Whether these surprise projects are to respond to opportunities, anticipate emergencies or restore destroyed systems and services, the lessons from the cases and other reports should help prepare to manage them.

Notes

This chapter draws upon some lessons to be learned previously stated in papers in the *Proceedings of the Institution of Civil Engineers*.

PART III
Future Needs

Chapter 8

And Now?

So What's Different?

The purpose of the case studies was to learn how managing urgent and unexpected projects is different from managing normal projects. The results indicate how businesses and government organizations could prepare to manage new business opportunities, crises and emergencies.

An obvious difference of urgent and unexpected work compared to normal is that rapid decisions are required on scope and planning the work with the resources immediately available. Planning what to do and how to do it proceed simultaneously. Established sequences of considering and evaluating choices before starting a project are compressed. Decisions on what to do when, how and by whom have to be made just as in managing normal projects, but made immediately.

Less obvious demands of urgency are the involvement of all levels of management and stakeholders, oral commitment to decisions and powerful project management to cope with the intensity of communications. If time overrides cost, the first decisions on what to do and how to do it have to be final decisions. To do this demands a unified team of expertise and authority combining all levels.

A standard model for managing urgent and unexpected projects does not emerge from the cases. The cases and the reports of other urgent and unexpected projects provide a variety of instances of how such projects were managed successfully. They are not necessarily representative of all possible situations. They do provide comparisons which may help prepare for such events. They illustrate how project teams were formed, stakeholders involved and resources deployed. They show choices for the management of such projects contingent on the conditions of urgency and uncertainty, the stakeholders' objectives, the scale of each project and the resources available.

Thoughts for Normal Projects

All these urgent and unexpected projects met their objectives. The reasons stated for their success were reviewed in Chapter 5. The same could contribute to the success of all projects, particularly these:

- Clear objective.

- Early recognition of managing work as a project.

- Dedication of full-time leadership.

- Involvement of stakeholders and the media.

- Close relationships between organizational levels.

- Dedication of resources.

- Concentration and continuity on the one project.

Conditions suggesting an emergency are commonly thought to be motivating and were sustained in several cases, even when the emergency had passed or was only predicted. The commitment of the public services volunteers to the 9/11 pile sifting work was sustained over months after any prospect of rescuing survivors. Few were personal stakeholders. Any other common challenge could be similarly motivating to the people facing it, and sustained.

Close relationships between all organizational levels in the cases may have been helped by the 'Hawthorne' effect of evident top interest and support (Handy 1987). The general lesson may be that motivation could be maintained by this support in normal situations where urgency is not a stimulus.

Dedication of resources in an integrated distinct team is not common normal practice in asset owners' or their suppliers' organizations. Dedication to a project team is recognized as motivating, but inflexible with resources, limited in learning and short term for careers. Instead, various forms of matrix structure are usual in all but small organizations. In them, continuity of work for any one project tends to be disturbed by the switching of resources from project to project according to various real or supposed priorities. On paper, employing people according to their specializations is logical, but managerially tends to emphasize input and its cost rather than output and its value.

Flexibility that is attractive managerially may be destructive psychologically. The general lesson for normal projects is to limit the switching of individuals' attention from project to project.

There were several such instances in the cases of managerial actions under urgency being closer to what their companies and advisers say should happen rather than what does happen 'normally' in their organizations. Why should this have been? Perhaps a surprise project was a fresh start, freed from its parent organization's habits. If the results were better than normal, why doesn't everybody manage all their projects in this way? And with first decisions agreed collectively and known to be final? All such expedients evolved under urgency should be reviewed for permanent adoption.

Reliance on oral commitment at meetings, particularly in the 9/11 pile case, was supported by discipline in clarifying decisions at the time that was remarked on as lacking in normal practice. Lack of that discipline is seen in business and other organizations in the farce of fuzzy decisions at meetings and the minutes of a meeting being drafted afterwards, circulated and then disputed, causing uncertainty and delay. These habits indicate lack of managerial attention to how the organization is being managed. Why not agree written minutes as you go, item by item, during the meeting, not after, as is the practice in contractual negotiations? As in achieving most improvements, this demands more time earlier to save much more time later.

Many of the problems in managing normal projects are a result of poor early decisions prior to the selection of a project. The explanation is that decisions are not made well initially because of the expectation of changes later. The cases showed that proceeding with no stage gates and no time to make changes forced successful first decisions. It may be that commitment to first decisions demanded better thinking than is common practice where later change is habitual. If this is generally true, an immediate start might be better for every project. Time is the greatest asset in projects, as in life. Cost is only more important when the completed project generates no measurable or significant value. A discipline of first decisions could be applied to all projects. In many Western business cultures this might take time to be effective, and this even more so in government and other non-commercial organizations.

Delays caused by changes to objectives, scope design and plan are often thought to be a problem in delivering normal projects to specification, on time, within budget and safely. Does knowing that first decisions have to be final on urgent and unexpected projects produce better first decisions than on 'normal' projects in organizations where later changes are expected? Whether optimism bias and changes may be tolerated as reflecting the real world or human weaknesses, they may be minimized by the concentration of the mind (or minds) characteristic of urgent and unexpected work.

Lessons of the Precipitating Events

Lessons learned in each case about anticipating the precipitating events are beyond the scope of this book, but several led to additions to systems for obtaining and acting on warning signals.

Whether worth it in terms of cost compared to probable value, systems may have to be made to at least try to avoid repetition of such events wherever the public are at risk. Insurers of private risks may tend to require the same. The consequent increase in the complexity of inspection, predictive and response systems may also add to the risks. Attempts to engineer risk need care to manage rather than fight Nature. Until an event causes loss, the common influence is that the preventive cost must show value. Carried to an extreme this demands that all events must be expectable. The result is a set of questions:

- Could and should condition reviewing predict any risk to an asset?

- Should we be surprised when external events damage an asset?

- Who sees the warnings?

- Who is listening for weak signals?

- Does trying to anticipate everything blind everybody?

These are questions for stakeholders and governments. All the cases were the result of surprises which afterwards might seem foreseeable. None of them were criticized at the time for not having been foreseen. And of course not all are adverse, for instance the building of the temporary rail station, the Freeview opportunity and the flood anticipation projects. An obvious lesson in risk management is that stakeholders should not only analyse their known unknowns, but should also know that they risk unknown unknowns and to prepare for them by drawing on the value of lessons from unexpected projects.

For the Future

All the cases were outliers, exceptions to normal practice. They prove no rules. They do provide precedents. The cases include some valuable information on *how* those projects were managed, as well as *what* was done, but as Flyvbjerg (2006) has commented about learning from reality, a few cases are a poor basis for general conclusions. The available cases of urgent and unexpected projects varied in their causes, the authorities involved, public concern, nature, scale, immediate resources available, uniqueness and location. They had in common that they demonstrated the potential value of developing the ability to manage surprises, what Weick and Sutcliffe (2001) called 'resilience'. Achieving success from surprises is increasingly vital, for instance to respond to unexpected business opportunities or the threats posed by rapid change in technology in the increasingly volatile commercial world. Or to respond urgently to overcome damage, whether imposed by human error, systems failure, Nature or terrorist attack. Any of these may be unexpected and urgent, and so demand an immediate start.

Adapting the words of the Association of Project Management (APM 2011) on corporate project governance, the provision and use of 'appropriate methods, resources and controls' for surprises should be agreed in advance. From studying losses due to natural hazards, Stretton (1979) concluded that false confidence too frequently resulted from belief that an emergency 'is something that happens to someone else'. That belief can reinforce reluctance to incur costs to prepare for unknown risks. Counter to that reluctance, Knight and Pretty (1996) showed that companies can gain in market value if thought to be able to manage crises.

To prepare for the unexpected, Weick and Sutcliffe concluded that organizations can develop managerial resilience by fostering a 'mindful' infrastructure which '[t]racks small failures, resists oversimplification, remains sensitive to operations, maintains capability for resilience and takes advantage of shifting locations of expertise'. This requires the capacity to hear and act on weak signals – the first signs of problems. In systems engineering this ability to detect the unexpected is rather unhelpfully called 'redundancy'. It's the organizational equivalent of the 'sixth sense' for human survival. It can be smothered in organizations which behave only as a collection of careers.

Weick and Sutcliffe drew on earlier studies of teams and organizations which had failed to make sense of unfamiliar threats and suffered 'disorientation, difficulty in comprehending, slowness of thinking, mental confusion, loss of ability to conceptualize alternatives or prioritize tasks' – as might be expected when suddenly faced with an urgent, unexpected project. The lessons from the reported cases are that to help anticipate these risks of disorientation, incomprehension or mental confusion every organization's culture should embed these principles for managing urgent, unexpected work:

- A vertically integrated team will decide priorities.

- Urgency will be defined with operational meaning.

- Specific communications channels will be defined.

- The team will include business continuity planners.

- Decisions will be made where the knowledge sits.

To develop resilience, managers should review how their organizations cope with minor events as lessons for coping with rare extremes. Attend to weak signals and establish an open culture to learn deliberately from small surprises as exercises to prepare for larger events. Every man-made or natural surprise could be used as a contribution to developing resilience and to take advantage of expertise to respond to both problems and opportunities which may need urgent and unexpected action. More often in practice, what are hoped to be improvements to practice are the reaction to problems. Yet the problems may be outliers. Remedies to rare problems considered alone could affect practice that is otherwise satisfactory. Organizations should plan to learn as much from what worked well as from what did not. Business continuity planners should guide and monitor this as part of their planning for emergencies. It is one of the few preparations possible for unexpected work.

Every year brings lessons from new surprises – known unknowns and unknown unknowns. Organizations need to be intelligent to respond. Resilient design and contingency planning could prepare infrastructure and services for man-made and natural surprises. In a study of achieving business continuity, Lienart (1996) concluded that a carefully thought-out plan made all the difference to relationships. Prior planning was valuable for understanding roles, resources and logistics and for establishing a common language for communications.

Institutional Lessons

The decisions said to have been crucial to their success, and specific lessons to be learned, were recorded from most of the cases. As with most experience, advance knowledge of these lessons could have led to earlier awareness of the differences from managing normal projects.

The arguments for preparing to manage through urgent, unexpected surprises are similar to the arguments of governments and insurers that businesses and other organizations should establish procedures to anticipate and manage crises and recover business continuity after major disruptions to their normal operations. The British Government-endorsed booklet PAS 200 (BSI 2011) on crisis management recommends that every organization should empower appropriate people to lead the development of the ability to manage crises. It recommends that roles and responsibilities should be established including that of the senior responsible owner, and the whole organization should be informed accordingly.

Such provisions for anticipating threats and crises could also be the understood basis for rapid response for new business opportunities. The advice on crisis management and preparing to ensure business continuity is written in terms of protection and recovery from sudden threats. What is needed to recover from disruption is no different to managing through any surprise that causes urgent unexpected work. The understood system for incident management observed amongst the twelve cases illustrated this preparation in some large organizations. They established strategic coordination combined with the expedient actions by the manager of the resources on the job.

As advised in PAS 200, surprise plans should provide a basis for an understood but flexible response. The plans should not focus on specific risks, as this may condition people to try to fit reality to expectation. The objective is to establish common understanding of how the organization will manage the surprise, but not fix what it will need to do. This planning should be only an addition to every organization's thinking about how it manages itself, the 'meta-management' function needed at corporate level in organizations observed by Beer (1972) as akin to the brain's hypothalamus function but in practice seen by Howard (1979) and Margerison and McCann (1984) to be lacking at many levels from business teams to British Government cabinets. When surprised, most if not all stakeholders and individuals are likely to give great attention to deciding what needs to be done. The sponsor is the one to give attention to how the team is working as well as what they are doing.

The common lesson is that organizations need what is deliberate governance that provides 'the structure through which the objectives of the project are set and the means of attaining these objectives are determined' (Turner 2006).

Preparing for Opportunities

All projects cause changes, not only by creating or changing a system, product, structure or service, but also in the careers of individuals and the collective experience of all parties. Projects can also be the vehicle to change organizations and their practices. *Carpe diem*, as said in one case. The changes in procedures reported from some cases to respond to urgency may be of lasting value. Loosemore (1998) in reviewing organizational behaviour during crises in construction cited an observation by Hermann that crises cause organizational and other changes. Improvements to assets and procedures have to be ready, agreed and usable for the surprise opportunity to apply them. The cases showed that surprise projects could be the opportunity for immediately helpful and longer-term changes, but only if the proposals for changes had been already thought through and agreed. Organizations need not be ill-prepared: changes can be prepared.

Planning for surprises can be valuable for understanding roles, resources and logistics and for establishing a common language for communications. As in planning for system maintenance, prepare through alternative plans, as any one plan or exercise simulation rarely fits events. Existing consultative structures between customers, public authorities, the emergency services and industry representatives established for planning for emergencies provided an effective basis for consultation on the work proposed in some of the cases, notably those where the urgent and unexpected project was an interim provision or addition to a long-term project already accepted as needed. That their senior staff had met regularly to plan for emergencies was an obvious advantage. Cultivating consultation and being aware of expertise available are two possible actions to prepare for unexpected demands.

Whether an event could or should have been expected can be a big issue to all affected by its consequences. The difference may seem irrelevant in managing actions to remedy the consequences, but it can affect preparedness and restrain actions.

Of course, information on urgent and unexpected projects which did not achieve their objectives is not so easy to find, perhaps because there is a natural reluctance to report failure. Tales of them circulate orally, largely within an industry, for instance of headless managerial actions that worsened a problem. Reliable lessons from them would be valuable. The tales indicate that some may have failed because they were started without agreement on objectives or without concentrated leadership, lacked recognition as projects and therefore lacked coherent attention to their objectives, choices and risks through to completion. This is reputed to happen among managers experienced only in operating existing systems. Without a common objective their expert resources will try to do what they know how to do. Their contribution to what is needed may be accidental.

Coiner of saws George Bernard Shaw is reputed to have said 'we learn from experience that we do not learn from experience'. Perhaps more accurately we do not learn first time or fast enough. Or the form of lessons is not attractive or the lessons are not recognized. Notable failures in engineering have been due to not asking questions (Byfield et al. 2008) or not asking them early enough. Lessons may be diffused among those with experience of events and those who extrapolate, but not drawn together for all. Many lessons from urgent and unexpected projects are there for those who look for them and potentially valuable for those who think about them. As Ellis (2012) wrote when summarizing lessons of crises, 'Be Prepared' is a mantra that can well serve individuals, companies, organizations and governments when seeking to manage urgent and unexpected events. The case studies, as well as other examples presented in this book, illustrate 'why' the effective management of urgent and unexpected events can be critical and have hopefully provided some lessons on the 'what, when and how'.

Appendix 1
Case Summaries

Notes

Several of the cases consist of work to restore systems and services after a disastrous event had caused severe damage. After such events the first actions are those of life saving, safety and investigation of causes, controlled by the police, emergency services and investigating authorities. These case studies start from when control of the asset is returned to its owner.

The Great Heck rail reinstatement and the Heck embankment stabilization projects were in the same area of the UK but were unconnected in their causes.

New TV Business, Freeview, UK, 2002

Scope

To design and launch the UK 'Freeview' digital television system in six months, May–November 2002, requiring an immediate start to technical development and consumer marketing to meet the criteria of the licensing authority and provide the basis for the successful launch and operation of Freeview.

Triggering Event

Financial collapse in March 2002 of ITV Digital and the return of three digital terrestrial television (DTT) multiplex licences to the regulatory authority, the Independent Television Commission (ITC). The multiplex licences were originally awarded by the ITC in 1998 as part of the development and operation of the UK digital terrestrial TV service.

Failure of ITV Digital had not been expected, though people in the industry had thought it was looking unstable.

The ITC re-advertised the three licences and allowed four weeks for the submission of applications. It was expected that the successful applicant would relaunch the digital terrestrial TV service within a matter of months in time for the Christmas market for buying viewing equipment.

The BBC formed a partnership with a transmission company and a commercial competitor to apply for the licences. In total, the ITC received four bids for the multiplex licences including the bid from the BBC and its partners.

Leading Stakeholders

The BBC initiated and led the project to bid for the licences, following the opportunity to implement government policy to increase the number of households with digital TV.

Timetable

On the failure of ITV Digital, the BBC's Director General initiated action to compete for the licences and gained immediate endorsement of Board members and its regulator, the Governors.

Winning the licences and then launching the product were the two stages of the project. These two stages were recognized from the start of organizing the management of the project.

30 April 2002	Previous licensee surrenders licences to UK television regulating authority.
I May	The authority re-advertises licences.
13 June	BBC and partners submit linked applications. Authority receives four bids.
4 July	The regulating authority announces BBC and partners have been successful in their bid.
30 October	Freeview service launched.

Project Organization

The Head of Projects in the Director General's office was Project Manager for the bid for the licence and Launch Director, reporting to the Director General. The Head of Business Management in BBC Distribution was the launch Project Manager.

The Head of Projects had experience of leading previous pan-BBC projects and had the full support of the Director General.

An early task for the Project Manager was obtaining agreement to an integrated project plan based on the three organizations' plans. This and establishing control was achieved through regular meetings, together with many informal discussions and socializing to achieve teamworking and manage cultural differences between technical and marketing staff.

This style of leadership followed the BBC Director General's policy and style of informal and open communication between all levels.

Resourcing

The project team was formed from existing staff of the three organizations with different expertise and different cultures. A single team was achieved through a combination of top management support, a flat structure, attention to team-building and dedicated project leadership.

The BBC operates as a TV programmes provider, not a transmitting organization. It took the lead on bidding for the licences, but for this formed a consortium with the company which owned and operated the terrestrial transmission platform and a company experienced in marketing commercial TV which could use capacity for its own free-to-air channels.

The resources needed were provided from the three partners' existing technical, marketing and management staff. Two were in the London area and one in Warwick. Because of the short duration of the project they continued to be based in their established locations. 'Virtual' working was established. This worked well. Once launched, the new system became a part of each partners' normal operations and a small operating company was established to undertake marketing and to manage the Freeview brand.

Performance

The project achieved its objectives. The launch was two weeks later than originally planned, but successfully met the target of attracting the Christmas market.

CRITICAL DECISIONS

- Taking the transmission company and competing commercial company as partners.

- Appointment of full-time Project Managers for the project and its launch.

- Appointment of a Project Manager able to coordinate staff in different companies and with different cultures.

- Agreement of all stakeholders to an integrated project plan.

- Injection of early control of the project, across all parties.

It was stated that reports that pessimists in the wider industry believed that the project was doomed to fail spurred the team to success.

Recorded Lessons

- Leadership by a small steering team and direct access of the Project Manager to the Director General.

- Operation of a 'flat' organization structure.

- Team socializing.

- Frequent meetings and open communication, creating 'a sense of interdependency, cooperation, trust and collaboration'.

Sources

BBC. 2002. http://www.bbc.co.uk/pressoffice/pressreleases/, last accessed October 2004.

Comptroller and Auditor General. 2004. *The BBC's Investment in Freeview*. National Audit Office.

Dyke, G. 2004. *Inside Story*. New York: HarperCollins.

Major Projects Association. 2003. *The Management of Unexpected Projects*. Seminar, London, June, summary published at http://www.majorprojects. org/pdf/seminarsummaries/108Unexpprojects.pdf, last accessed 3 April 2014.

Appendix 1.2

Temporary Rail Station, Workington, 2009

Scope

Construction of a temporary new rail station including two platforms, footbridge, waiting room and gravel car park in the northern part of Workington to enable Cumbrian Coast Line trains to provide a passenger service reconnecting both sides of the town after the collapse or closure of the town's road bridges over the River Derwent. The platform structures and footbridge were constructed of scaffolding covered with wooden planks with an anti-slip surfacing. After first completion, the length of the platforms was increased for the longer trains needed for the volume of passengers using the service, and the area of the car park was extended to a total of 500 places.

Precipitating Event

Sustained intense rain storms over much of western Cumbria had resulted in severe flooding of the River Derwent causing collapse or closure of both road bridges linking the northern part of Workington and the main area of the town. As a result, road travel between the two was a 30 mile journey.

The rail bridge over the Derwent remained undamaged so that the train services that connected Workington and stations northwards could continue, but the next station north of the river (at Flimby) was four miles up the line from the isolated part of Workington.

Timetable

20 November 2009	Extreme flood levels destroyed Workington's Northside road bridge over the River Derwent.
22 November	Safety closure of the Calva Bridge, the town's last surviving road bridge. Northern Rail proposed construction of the 'Workington North' temporary rail station.
23 November	All parties agreed the 'Workington North' temporary rail station project was feasible.
24 November	Construction began during the night of 24/25 November.
26 November	Southbound platform completed.
28 November	Northbound platform and footbridge completed.
30 November	Workington North station opened. Additional shuttle train services added to existing train services.
2 December	Temporary car-parking area completed.
21 April 2010	Replacement road bridge over Derwent opened.
28 May	Shuttle train service ceased.
8 October	Temporary station closed. Temporary station removed and car park restored.

Leading Stakeholders

Northern Rail, the train operating company (TOC) of the scheduled Cumbrian Coast Line services from, to and through Workington, proposed the construction of a temporary Workington North station.

Network Rail plc, the owner and controller of the track, agreed to construct the temporary station and car park, at their cost.

Allerdale Borough Council gave planning approval for the temporary station and car park, leased the land and provided the power source for the platform and car-park lighting.

The Cumbria Flood Recovery Coordinating Group formed by the Cumbria County Council, the local councils, police and the emergency services coordinated all parties.

The UK Department of Transport met the cost of operating the trains between Workington and the temporary Workington North station as a free shuttle service until the completion of a replacement road bridge.

Direct Rail Services provided additional train services from Maryport through to Workington to cope with the demand.

Construction of a temporary footbridge between the main and north parts of the town by the Royal Engineers was at a separate site clear of the construction of the temporary rail station.

Project Organization

Network Rail's Property Group based in Stockport was made responsible for the planning and construction of the Workington North temporary station project.

A geotechnical engineer already on site to check the effects of the river floods on the rail bridge moved on to assess the site for the temporary station. Network Rail's helicopter was patrolling in Cumbria on 23 November and was dispatched to the proposed site to provide aerial footage to assess the site.

A three-man Network Rail team was seconded full time to lead the project:

- Project Manager – the Property Works Manager, reporting to Network Rail, Northern, Allerdale Council and the other stakeholders.

- Contracts Manager.

- Works Delivery Manager.

Their normal duties were undertaken by deputies. It happened that deputies for two had already been arranged for absence for training.

Design of the temporary platforms was undertaken by Network Rail with the contractor. Starting from a temporary platform detail and footbridge design used previously in Network Rail during track maintenance allowed work to commence almost immediately following assessment of ground conditions at the chosen site.

The construction of the platforms and footbridge was supervised by a Chartered Engineer and changes to design made on site as needed. An independent structural engineer was employed by the contractor to check the safety of the scaffold design and construction. The completed work was checked by Network Rail engineers prior to acceptance.

All worked 12+ hour days on site. At this time of year, early mornings and evenings were in darkness. Up to 40 people were employed on site, at the maximum on the Friday.

Local businesses provided secure storage yards.

The local community provided tea, coffee and other help.

The trackside work was subject to all normal safety requirements.

The project team held conference calls twice daily to review and report. The project manager ran daily conference calls with all stakeholders.

Procurement

The day it was agreed to proceed with the temporary station, two pre-qualified contractors usually employed to supply and erect scaffolding were invited to offer competitive tenders to construct the two platforms and passenger footbridge. The one chosen was able to offer immediate use of material that happened to be already loaded on trucks following recent completion of other work. The programme in the contractor's method statement showed the sequence of work in detail. The terms of payment were fixed price for construction plus hire time rates.

For the ground works a contractor was deployed by the Network Rail Carlisle maintenance depot under an existing framework contract.

Electrical services were the work of an approved framework contractor.

Local firms were employed to supply the platform coverings and car park gravel.

All access for materials and equipment was by road so that normal train services would not be interrupted.

Supervision and safety checks were the responsibility of the Network Rail Property Services team.

Performance

The southbound platform for the temporary station was completed within two days, the northbound platform and footbridge two days later, as planned.

The existing train services continued throughout construction of the temporary station.

There were no reportable accidents on site during the work.

Though heavily loaded with flood rescue and recovery work over the whole county, the local authorities provided all the support and approvals requested by Network Rail and their contractors.

The temporary rail station and car park were removed and the area restored after a replacement road bridge over the Derwent was opened in April 2010.

The project established a design of temporary platform for future needs.

Direct Rail Services together with Northern Rail, Network Rail and local authorities won the Outstanding Teamwork Award at the 2010 National Rail Awards.

Recorded Lessons

The project team considered that the keys to the success of the project were:

- Network Rail's Property Groups were used to emergencies and other reactive work and had contractors in place for immediate calls.

- Temporary footbridges are often required during track work and so a basis for design and pre-qualified contractors were available.

- Good cooperation between all parties. Communications on the project were good as many individuals in Northern Rail, Network Rail, the contractors and subcontractors were used to working together.

Sources

Building a train station in a week after Cumbria floods. 2009. BBC News Online, 25 November 2009, http://news.bbc.co.uk/1/hi/uk/8378702.stm, last accessed 3 April 2014.

Flood town of Workington is cut in two as bridges collapse. *The Sunday Times*, 22 November 2009.

Livesey, C. 2011. The Barker Crossing: Royal Engineers reconnect Workington. *Proceedings of the Institution of Civil Engineers*, 164(CE2), May, 81–7.

Rail station hope for the town cut in two. 2009. BBC News Online, 24 November 2009, http://news.bbc.co.uk/1/hi/uk/8377599.stm, last accessed 3 April 2014.

Teamwork Award, National Rail Awards, 2010.

Workington North temporary station progressing well. 2009. Network Rail, 29 November 2009, http://www.webcitation.org/5leXFTx2i, last accessed 3 April 2014.

Acknowledgements

Thanks are due to Network Rail plc. for information, advice and comments on drafts for this case summary.

Appendix 1.3

Thames Bank Raising, London, 1971

Scope

The project required raising and strengthening 38 miles of the Thames tidal river banks and linked structures within six months. The construction work to be done had to suit the state of the banks and other defence structures in over 800 properties including private houses and gardens, parks, highways, historic buildings, offices, warehouses, industrial yards, wharves, docks and public utilities. Many of these properties were in daily use, but some were abandoned, derelict and hazardous.

The project was not large in total cost compared to the range of civil engineering works commonly undertaken by a metropolitan authority. It was unusual in being urgent and unexpected and in consisting of many relatively small concurrent operations to be carried out in properties mostly not controlled by the responsible authority.

Triggering Event

London needed increased protection against high Thames estuary water levels during North Sea storm surges and spring tides as the south-eastern part of the UK is slowly sinking and relative sea levels are rising. It was fortunate not to be severely flooded in 1953 when the Essex coast in south-eastern England and, on the other side of the North Sea, stretches of the Netherlands coastline experienced fatal breaching. Late in 1970 the UK government and the Greater London Council (GLC) agreed that greater defence should be provided by constructing a tidal barrier of gates lower in the Thames estuary. Completion of that project was expected to take six years. To provide some temporary additional protection until then, the authorities also agreed to proceed with the Thames Interim Bank Raising project to increase the flood defence levels by 18 inches as a matter of urgency.

Leading Stakeholders

The GLC was then the statutory authority responsible for control over developments by riparian owners to ensure that frontagers provided and maintained river defences. The Ministry of Agriculture, Fisheries and Food (MAFF) was the sponsoring government department.

The many owners and users of the riverside properties were directly affected by the need for access and construction work. Property owners and inhabitants of the surrounding lower-lying areas dependent on the river banks for flood protection were indirect stakeholders. These stakeholders included government and business offices and a large proportion of the GLC electorate.

Construction proceeded under powers to the GLC under the Land Drainage Acts. The work was subject to the consent of MAFF and consultation with local planning and other statutory bodies.

The cost was shared between the GLC and the government, through MAFF.

Timetable

December 1970	Decision to proceed with the Interim Bank Raising project. Formation of project team.
February 1971	Establishment of site staff. Tendering by contractors.
March	Five bank-raising contracts let.
April	Sixth bank-raising contract let.
May	Central area Embankments-raising contract let.
December	Work completed on one contract.
January–April 1972	Work completed on all other contracts.

Project Organization

The GLC Department of Public Health Engineering (DPHE) was responsible for the planning and execution of the project and for coordination with the frontagers, other GLC departments and the London boroughs. An exception was that the GLC Department of Planning and Transportation (DPT) was responsible for the simultaneous similar work required on the central London embankments.

Following the decision to proceed with the project, the DPHE established a temporary organization to be responsible for its management and design.

The Project Manager was appointed under a system established in the GLC for the management of projects 'of some size' and dependent upon inter-departmental collaboration. Under written terms of appointment the Project Manager was responsible for calling together a project team consisting of representatives of the GLC departments involved to define and coordinate all work for the project. The person appointed Project Manager was an experienced civil engineer who had been employed on the preliminary work for the project.

Area Managers were appointed for each construction contract for a section of the lengths of the banks, and assisted by other engineers, trainees and clerks of works. These were temporarily employed on secondment from firms of consulting engineers. They became responsible for on-the-spot negotiations with property owners and occupiers, for making design decisions, obtaining the agreement of all others concerned, for instructing the contractors and for supervising their work. Coordination of their design standards was provided by criteria set by the section in the DPHE established to check the maintenance of the existing flood defences.

A Deputy Project Manager was appointed and led a small staff of engineers and assistants chiefly employed on the planning and administration of the contracts.

A project coordination team was formed consisting of representatives of the other GLC departments (transportation, housing, public health, legal, etc.) concerned with the project. The team met formally during the first months after the decision to proceed with constructing the project. These meetings were the means of resolving doubts or differences about responsibilities for decisions. Initially the formal meetings were called more-or-less weekly and then less frequently as procedures for making decisions became understood.

Under the system established in the GLC for inter-departmental projects, the Project Manager reported to a Steering Committee consisting of the heads of all departments concerned. The Project Manager was called to meet the

Committee after his appointment. He reported to the chairman before its subsequent meetings, but would attend only if a difficulty on his project was to be discussed.

Resourcing

Public works contractors were invited to tender to carry out the construction work, as the DPHE did not employ directly the resources of labour and supervision needed for such a project. As the work was to be done in properties stretching along the two banks of the river, each bank was divided into sections logistically practicable for contractors.

The DPHE had little information on the structural state of many of the properties, the owners and occupiers having been legally responsible for maintaining the previous line of defence against flooding. Estimates of the type and extent of work needed in many properties were obtained by observation from boats, to obtain rapid if not accurate information and minimize disturbance to the property owners and users. From this, twelve types of bank-raising work were expected to be needed and were specified as the basis for contracts. As information became available from access to properties, the Area Managers decided the design and construction detail appropriate to each property and acceptable to all concerned.

Incentive and target types of contract can be appropriate for urgent projects, but they were not frequently used in municipal engineering in the UK. Getting agreement to use an unfamiliar type of contract might have taken too long. The 'Bill of Quantities' type of contract following common British practice was chosen for most of the work, modified in some clauses to overcome the initial uncertainty of knowledge of the structural state of many properties. It was known that, during construction, variations to designs were required as the real state and needs of properties were explored, but the choice of this type of contract enabled contractors to be chosen rapidly on the basis of competitive tendering, and construction to be started to meet the urgent programme for the project.

Contractors were selected who were already approved by the DPHE for the type and amount of civil engineering and building work in each section. Following GLC standing orders, a minimum of four tenders were required for each contract. The larger contractors were permitted to tender for all six contracts. One was awarded three of the contracts, others one each.

The DPHE Director of Engineering was 'the Engineer' responsible for the administration of these contracts following conventional British practice.

Performance

The bank raising and related work were completed to programme. They provided the required protection until the flood risk was overtaken by the protection provided by the permanent barrier.

The major risk that had been expected was delay in obtaining access to properties and agreement on the work needed. This was minimized by delegation of all communications with property occupiers and decisions on the work appropriate to their properties to the Area Managers. In the event, all occupiers and owners agreed to allow access for the work. The GLC's legal powers to enter properties were never needed.

Decisions on the bank raising and strengthening work to be done in each property depended upon identifying and contacting owners and occupiers, arranging for access to each property, agreeing design proposals with them and other interested parties, and making the arrangements for the construction work to be done. Written advice of the need to do so was first sent to owners and occupiers, followed by calling to agree what to do when a contractor was ready to proceed. To minimize disturbance to occupiers the area team were prepared to make a single visit to a property to inspect the bank, decide what work was needed and arrange for their contractor to have access.

The contract specification of twelve types of work provided the Area Managers with the flexibility to use whichever was suitable for each property and acceptable to all parties. Pre-agreed designs and monitoring of their application anticipated the risk of inconsistent standards of protection.

All potential problems between GLC departments were resolved by the project team without calling on the higher-level role of the Steering Committee.

Recorded Lessons

The following were agreed by the project team as the keys to the success of the project:

- Project Manager appointed under accepted system for inter-departmental projects.

- Prior briefing of property occupiers and owners.

- Personal contacting of occupiers and owners wherever possible.

- Area managers authorized to take the lead with occupiers and owners.

- Once-only disturbance of occupiers.

- Division of the construction work into sections suitable for the contractors.

Delay in proceeding with the project was minimized by:

- Following established procedures in forming the project team.

- Inter-departmental cooperation.

- Using established terms of construction contract and tendering procedures.

- Providing alternative approved designs to meet the needs of different properties.

- Employing already known and approved contractors.

Sources

Gilbert, S. and Horner, R.W. 1984. *The Thames Barrier*. London: Thomas Telford.

Greater London Council. 1971. *Thames Flood Prevention – Thames Barrier Project*. Second report of studies.

Rothwell, D., Thompson, P.A. and Wearne, S.H. 1975. *Management of an Urgent Public Works Project*. Report, School of Technological Management, University of Bradford, and Public Works Division, University of Manchester Institute of Science and Technology.

Rothwell, D., Thompson, P.A. and Wearne, S.H. 1976. *Management of an Urgent Public Works Project*. International CIB symposium, Washington.

Schofield, A.N. 1970. Cost of preventing Thames tidal flood. *Nature*, 227(5264), 19 September, 1203–4.

Wearne, S.H. 1972. Project management along the Thames. *Civil Engineering and Public Works Review*, August, 813–14.

Acknowledgements

Thanks are due to the Greater London Council and the Director of their Department of Public Health Engineering for access to the work and to documents, and to members of the project team interviewed for their time and their comments on draft reports.

Postscript

The permanent Thames flood barrier was completed in 1982. Much of the higher river banks constructed in the interim project were left in place.

Appendix 1.4

Flood Diversion Scheme, Chichester, 2000

Scope

To design and construct in two weeks an emergency river flood diversion system of pumping, piping, temporary channels and culverts around Chichester, in total 12 km in length, December 2000.

Triggering Event

Widespread and large-scale rainfall in autumn 2000 caused extensive flooding and surcharged ground water aquifers leading to predictions that flow in the River Lavant through Chichester would rise to 17 times the long-term average. The project was required to divert the predicted excess flow.

Following flooding in the city in 1994 the need for a flood diversion scheme had been accepted by city, regional and national authorities and a permanent scheme had been agreed for construction starting in 2001. The need for the emergency project in the autumn of 2000 was due to the most widespread and large-scale flooding across England and Wales for over 200 years. The emergency project was an urgent and mainly temporary version of the agreed permanent scheme.

The emergency diversion was urgent because fully charged aquifers north of Chichester could have siphoned a sudden additional flow down the river. In the event, this happened. The resulting peak flow arrived a week after completion of the emergency diversion.

Leading Stakeholders

Following the flooding in 1994, the Strategic ('Gold') level of public emergency planning as defined in the Home Office guide *Dealing with Disaster* had been established by the West Sussex County Council to provide a framework of consultation and policy on flood protection between all responsible authorities. Gold control brought together the County Council, Chichester District Council, the Government's Environment Agency, West Sussex Fire Brigade and the Sussex Police at chief executive and equivalent level. Gold met annually in September to review threats in the coming winter.

With sustained rainfall in the autumn of 2000, the Agency reported to Gold in October 2000 that a temporary flood alleviation scheme might be required. A Tactical ('Silver') level of control consisting of representatives of the operational levels of all the above and industry was therefore established to plan all the options for the emergency work, coordinate priorities in allocating resources, obtain other resources as required and coordinate construction requirements. A consultative group was also established with the users of an industrial estate which would be one of the first places to suffer a repeat flood. The Gold level of control remained responsible for monitoring overall problems and arrangements for military aid to stand by in case the actual floods were seriously in excess of the combined capacity of the culverts, pumping through the city and the temporary diversion scheme.

On agreeing the need for the emergency scheme the County Council and the Agency arranged that its construction should be undertaken under the Agency's emergency powers, with the County Council acting as agent for the Agency to manage design, construction and public relations. By following the line of the planned permanent diversion scheme the emergency project did not need additional statutory consents. Rapid agreement for access to land was facilitated by the relationships established for the permanent scheme.

The 1994 floods had threatened a mainline railway line. Discussions with Railtrack on the culverts needed for the permanent scheme helped with agreement to the emergency work, but with this a different design of culvert and method of construction were proposed in order to achieve faster construction.

The costs of works not usable for the permanent scheme were met by the County Council, to be reclaimed from central government under rules for emergency work. The costs of works which were usable for the permanent scheme were met by the Environment Agency, on behalf of the local flood defence committee, with grant aid from central government, following the usual rules for work on main rivers.

The out-turn cost of the emergency project was higher than estimated as the opportunity was taken to construct parts of the permanent project where this could make best longer-term value.

Timetable

The decision to proceed with the emergency project was made immediately by the Agency and County Council as the need to do so had been established through their link at Gold level of consultation and planning.

Project Stages:

1.   Decision to proceed with the emergency project.

2.   Decisions on the line and design of the emergency work.

3.   The construction work, in three simultaneous sections.

28 October 2000	Emergency pumping through city put in place for 0.5 cumec (cubic metres per second).
6 November	River flow from groundwater and rainfall up to danger level. Emergency pumping through city increased to 1 cumec. Need for additional flood flow diversion predicted.
9 November	River flow up to 1994 level.
14 November	Design and consultations for emergency project started.
1 December	Strategic 'Gold' control authorize project.
4 December	First joint meeting with contractors.
5 December	Emergency river flow diversion started into pits.
6 December	New channel setting out and archaeological work started.
11 December	Contractors start work.
15 December	River reaches peak flow of 8.5 cumec.
16 December	Emergency river flow diversion operational.
5–7 Jan 2001	Possession for constructing permanent culverts under railway.

Project Organization

The emergency project was able to draw on the professional services of the County and Agency departments and their consultants already employed for the permanent scheme due to start construction the next year.

The County Council and the Agency appointed Project Managers to work in tandem reporting to Silver control. The Agency's Project Manager was the person already in post to manage the permanent scheme and thus familiar with the design and details of the project and, importantly, all the interested parties. Once decisions had been made at Gold level to authorize the work, the Project Managers were left free to direct the work for the project. They reported to Silver level on progress and through them obtained high-level support if needed for solutions to problems. On behalf of the whole team the County Council's Project Manager reported to Gold control.

The temporary team needed was thus formed by dedicating to the emergency project staff already established for the permanent scheme and therefore known to each other and with contacts established with other interested parties. For the project, the Project Managers were the link between Gold, Silver, consulting engineers and all parties.

At the start of the construction work all parties, including the construction contractors, were invited to a briefing on the plan and responsibilities for the work.

Resourcing

Consulting engineers were employed by the County Council for design of the emergency project, the preparation of contracts and the supervision of contractors, reporting to the County Council's Project Manager. Consultants already employed by the Agency for the design of the permanent scheme provided the Agency's Project Manager with specialist advice. County Council and Agency departments provided the supporting legal and other services required. The Agency's Project Manager was responsible for approving the design on behalf of the Agency to ensure that it was compatible with the permanent scheme, and for liaison with the Agency's other departments, landowners, local planning authorities and statutory consultees.

The linear form of the work made it possible to divide the construction of the emergency works into independent sections. It was therefore divided into three contracts each appropriate in size for the resources available from three local contractors. One contract was for the intensive work for the railway culverts, to be carried out over one weekend. The other two contracts were mainly for the works north and the works south of the railway.

The contracts were based upon the ICE Minor Works Conditions of Contract, on a cost-plus basis of payment. The consulting engineers were responsible for supervision and for accurate and detailed record keeping, to provide the basis for payments to the contractors.

The estimated construction cost of the contracts was below the threshold for them to have to be announced in the Official Journal of the European Community.

Contractors were selected who were known, had the appropriate resources and could make a quick start. The contracts were made by exchanges of letters.

In response to the urgency of the work, the contractors worked seven days a week, with 24-hour working at some locations to minimize disruption to the public.

The use of temporary piping and pumping was coordinated by the Fire Brigade. Pipes for this were obtained from all over the UK and the rest of Europe.

Performance

Because of the urgency, changes were kept to a minimum. As noted above, the Project Managers allowed changes to design to facilitate construction.

The emergency project met all its objectives. A week after its completion the flood flow in the Lavant rose to the peak amount predicted. The emergency diversion, together with the pumping and other protective actions in the city, succeeded in averting extensive flooding to the city. Without it, 1,000 or more properties would have been flooded, as in 1994, and it is estimated that the damage would have cost over £8 million (at then current prices).

The emergency project was planned and authorized within a period of six days. It was carried out during a period of many other demands on Environment Agency and County Council staff, their consultants and contractors and all authorities over the whole area for flood control and many other measures for flood protection.

Long hours, night working, difficult (December) conditions and poor weather all provided possible site risks. A local consultant was therefore appointed as the Planning Supervisor as required under the UK Construction (Design and Management) Regulations for the safe design and construction of the emergency scheme. The consultant with this role for the permanent project was not available and not located locally. There were no reportable accidents during construction.

CRITICAL DECISIONS

- Agreement to proceed with the emergency scheme when need was uncertain.

- Quick decisions on the line for the emergency project based upon knowledge of the ground built up for the permanent scheme.

- Early establishment of project management.

- Appointment of the project teams already familiar with all parties and the need.

- Employment of consultants already experienced in the area.

- Employment of known contractors with the resources required.

- Use of appropriate standard Conditions of Contract.

- Agreement on cost-plus terms of payment.

- Detailed direction and control by the project management team.

- Round-the-clock working by project team, consultants and contractors.

Recorded Lessons

- All parties linked by dedicated project managers.

- Project managers' authority seen to be given by Gold level.

- Project managers immediately effective because of knowledge of the ground and planning for the permanent scheme.

- Cooperative attitudes established at the start of the construction work through all parties, including contractors, attending a briefing on the plan and responsibilities for the work.

- Local construction safety supervisor valuable to cover round-the-clock working.

Note

The permanent flood diversion scheme was completed two years later as planned and replaced the temporary sections of the emergency project.

Sources

Environment Agency. 2000. *River Lavant Flood Alleviation Scheme*. Environment Agency, Southern Region.

Environment Agency. 2001. *Autumn 2000 Floods Review*. Regional Report, Environment Agency, Southern Region.

Gilham, A.M. 2001. *River Lavant Flood Alleviation Scheme*. Chartered Professional Interview report, Institution of Civil Engineers.

Greeman, A. 2002. Micro diversion. *New Civil Engineer*, 31 January 2002, 24–5.

Hoad, R.S., Gilham, A.M. and Fawcett, D.S. 2003. Chichester emergency flood alleviation project, winter 2000/2001. *Proceedings of the Institution of Civil Engineers, Water and Maritime Engineering*, December, WM4, 297–304.

Home Office. 1997. *Dealing with Disaster*.

Appendix 1.5

Ouse Banks Heightening, Selby and Barlby, Yorkshire, 2000

Scope

Urgent replacement of emergency sandbagging with sheet piling along long lengths of the UK River Ouse banks after heavy flood flow when the bank was waterlogged and overtopping.

- Drive sheet piling deep into waterlogged river banks over a distance of approximately 4.5 km.

- Strengthen some sandbagging, remove other sandbags and clean up.

- Safety works to piling.

The scope was increased during the work to revise a section of piles and add wood capping along the heads of the piles for public safety.

Triggering Event

September 2000 saw twice the normal rainfall over south Yorkshire and heavy rain throughout October saturated much of the region. A further series of low pressure weather systems at the end of October caused heavy rain resulting in extensive flooding. Over 250 mm rain fell in 12 days. Rivers in the region rose to the highest levels ever recorded, causing extensive flooding in some areas. One result was overtopping of the River Ouse flood defences at Barlby, where 152 properties were flooded.

A massive deployment of sandbags (some via army Chinook helicopter) on the left bank of the river from above Barlby and down to Selby had temporarily prevented overtopping, but piping through and scour caused flooding in Barlby and seriously threatened the stability of the bank. At one time the river level was about 300 mm above the level of the bank and was held by the sandbags. Sheet piling was therefore required urgently to replace the sandbags and strengthen the bank. Some piling was also required on the right bank. That work was suspended at one time due to a widespread Foot and Mouth epidemic.

Leading Stakeholders

For the purposes of flood defence the Yorkshire Ouse is classified as a main river. For main rivers the Environment Agency inter alia issue flood warnings, advise national and local government and other parties on actions needed for flood control and implement approved projects. The Agency had permissive powers to do work on assets. The Agency therefore undertook the planning and execution of the emergency work at Selby and Barlby.

Tactical ('Silver') level of emergency control was opened by Selby District Council with the Environment Agency, and then coordinated by the Police.

The cost of the emergency work was met from Exchequer funding administered by the Ministry of Agriculture, Food and Fisheries (now the Department of the Environment, Food and Rural Affairs).

Timetable

29 October–6 November 2000	Three separate heavy rainfall peaks over the region.
30 October	Incident Room opened in the area to manage the Agency's response jointly with other authorities.
2 November	Tactical ('Silver') level of emergency control opened in Selby.
2–6 November	Sandbagging along river bank.
25 November	Environment Agency appoints its Combined Capital Works (Phase 2) (CCW2) team to plan and manage the major emergency work needed.
27 November	Scope and allocation of emergency work defined at CCW2 meeting.
mid-January 2001	Start of piling work.
mid-February	Access restricted by Foot and Mouth controls.
18 February	Piling completed except where restricted by Foot and Mouth controls.
12 April	Sandbags strengthened or removed.
18 May	Pile capping and clearing site completed.

The priority at Barlby was to complete the piling to ensure the protection of the village from the river before flood levels which might recur in the winter.

Some of the work planned for the CCW2 contractor and consultant team was already affected by the wet weather and consequent floods, and so was not delayed by giving priority to the emergency work.

Access to the river banks was interrupted in mid-February for eight weeks by the Foot and Mouth restrictions before controls were agreed to allow work to continue.

Project Organization

The Agency's Operations Department in the area includes resources for normal maintenance of flood defences on main rivers. By October these resources were being employed round the clock on sandbagging and other emergency work. The temporary sandbagging at Selby and Barlby was carried out by them with the assistance of local labour and army support. These resources were not available or suitable for the piling and other emergency work at Selby, Barlby and elsewhere. The Agency therefore appointed its CCW2 design and construction team to assist with the major emergency work needed over the whole region.

The CCW2 team had then only recently been appointed by the Agency under its National Capital Works Programme to design and construct a planned programme of river and flood defence improvements in the region, under the specific direction of the Agency. The CCW2 team consisted of a major contractor and engineering consultants. This established team provided the resource to commence assessing the needs at Selby and Barlby. The contractor and consultants then formed a small supplementary team for the emergency work for Gowdall and then for Selby and Barlby.

The CCW2 team reported to the Agency's Project Manager, National Capital Works Programme, as for their planned programme of work.

For the emergency work the Project Manager reported to the Agency's Regional Capital Works Manager and the Area Flood Defence Manager.

Procurement

The CCW2 contractor and consultant team was responsible for design and construction of the piling and capping. The contractor was employed under the New Engineering and Construction Contract. For the Barlby emergency work payment was cost-plus, under Option E. The contractor employed the consultants under the same terms of contract.

All payments to the CCW2 team for the emergency work were settled.

Normal procedures under the Construction Design and Management (CDM) Regulations were followed. The Planning Supervisor was already in place for the CCW2 contract. A health and safety plan and method statements were compiled and approved.

Attention was given to good lighting of the banks and the access to them.

Performance

The piling was completed before the risk of high tides.

There were no reportable accidents during the work.

The work was carried out during a period of many demands on Environment Agency staff, their consultants and contractors and all authorities over the whole region for flood control and many other measures for flood protection.

It is estimated that the temporary work prevented flooding of some 8,000 properties.

Recorded Lessons

The success of the emergency work critically depended on:

- Strong cooperation between the Environment Agency, local authority, police, fire and rescue services during and immediately following the flood event.

- Professional competence of all staff.

- Use of the contract provisions pre-arranged for planned works.

- Use of a national procurement contract to obtain a large quantity of sheet piles in a short timescale.

- Competence and willingness of consultants and contractors.

Sources

Environment Agency. 2001. *Autumn 2000 Floods Review*. Regional Report, North East Region.

Harman, J. et al. 2002. Floods – A new approach. *Proceedings of the Institution of Civil Engineers – Civil Engineering*, 150, May, 2–59.

Lane, G. et al. 2001. *36th Conference of River and Coastal Engineers*. Keele University, London: Department of the Environment, Food and Rural Affairs.

Appendix 1.6

Motorway Viaduct Propping, North London, 2011

Scope

To support temporarily a 10 m length of 36 fire-damaged pre-stressed concrete main beams of the M1 motorway viaduct at Dean's Brook, North London, approximately 5 m above ground level, by installing an array of Superprops, spreaders and smaller props in order to permit partial resumption of motorway traffic the next day, together with the traffic management of the motorway and connecting links and stand-by provisions for some motorway counter-flow.

Precipitating Event

The damage was caused by a fire in a scrapyard under the viaduct, with combustion of heavy rubber tyres and other industrial waste which caused extensive surface damage to the undersides of the pre-stressed concrete beams.

Timetable of Propping Project

Friday 15 April 2011	
0425	Fire under the viaduct reported by highway patrol.
0430	CPS Emergency manager (Silver) and structures engineer advised.
~0500	Section of M1 closed. Structures engineer on site but excluded from fire risk zone.
0600	On-call civil engineering contractor mobilized.
0800	First project management meeting at CPS. Significant damage to southbound carriageway identified.
1030	Fire Brigade allow access to site. Significant damage to northbound carriageway identified.

1100	Propping contractor at CPS. Civil engineering contractor commences clearance of fire debris, etc. and ground investigations.
~1200	Local residents, services and hotel appraised of situation.
1400	Propping contractor offers preliminary design. Temporary works design started.
1600	Propping contractor commences loading material for transport to site.
1700	Civil engineering consultants briefed for permanent repair work.
1800	Propping contractor's initial design submitted to CPS.
Midnight	Propping contractor starts transporting material to site.
1300	Propping commenced under northbound carriageways working west.
1834	One lane northbound (hard shoulder) opened to traffic*
0930	Additional lane northbound opened to traffic.* Propping starts under southbound carriageways.
1613	Two lanes southbound opened to traffic.*
1100	Need for additional propping under southbound carriageway identified.
1800	Additional propping delivered to site.
1100	Propping completed under three lanes both ways. Preparations for three lanes both ways re-opening.
0510	Three lanes running southbound.*
0620	Three lanes running northbound.*

Note: * with 50 mph speed restriction until the permanent repair of the beams completed in October 2011.

Leading Stakeholders

The Highways Agency (HA) is the UK government's body responsible for maintaining and developing England's road network.

Connect Plus Services (CPS), a consortium comprising Balfour Beatty, Skanska, Atkins and Egis Road Operation UK, is responsible under a DBFO (Design–Build–Finance–Operate) contract with the Highways Agency for the management and maintenance of the M25 system and some linked roads, including this section of the M1 motorway.

Local residents, services and hotel were appraised of the situation orally from midday on the Friday. Press releases and a letter drop to residents followed on the Saturday afternoon.

Project Organization

CPS were responsible for planning and managing the temporary propping project, and traffic management and diversions under their contract with the HA. The CPS Control Centre provided the communications links for traffic matters.

The London Fire Brigade established Silver level emergency coordination of the police, HA and CPS, for the fire, closure of the section of the motorway, the propping project and traffic management.

The CPS propping project management team consisted of:

- CPS Service Delivery Director – team manager and responsible to the HA, Silver control and for public relations.

- CPS Structures Engineer – responsible for directing propping contract.

- CPS Site Manager – site coordination of propping project plus back-up staff during night breaks.

The team was based at a CPS depot immediately north of the site.

Site safety – the CPS Service Delivery Director acted as the Principal Contractor.

Costs were recorded against the call-off contracts.

Nine conference calls held over the four days were the chief means of linking the HA and the CPS teams concerned with the propping, traffic management and other work.

All members of the project team were dedicated full time to the project until their roles were completed.

Press briefings were run by the HA. A 15-minute break was arranged in the propping work in order that the Press could enter to see the damage and appreciate the tight conditions of the work.

Resourcing

The civil engineering contractor and the propping contractor were employed under Variation Orders to framework on-call contracts established for small work and maintenance.

The terms of payment for the supply and installation of the propping were established rates.

Performance

The project achieved its objectives of staged resumption of northbound traffic over the damaged section of the motorway before the expected Saturday evening peak.

The project team stated that the critical decisions for delivering the propping project were:

- HA, CPS, police and emergency services had well-established links for daily traffic management and emergency planning.

- CPS quick response for analysis of the potential damage.

- The availability of contractors under framework contracts.

- Priority given to achieving partial re-opening of northbound M1 lanes: Access for propping under the viaduct was possible only from the west side. Priority was given to restoring two northbound lanes for football supporter traffic expected to be passing on Saturday at about 8 pm. As a consequence of this priority the speed of the subsequent work of propping the southbound lanes was limited because access was possible only between the completed northbound lane props.

- Rapid response of propping contractor and availability of their resources.

- Re-opening time not promised too soon.

There were no lost time accidents during the project.

The cost was met by CPS's insurance.

Recorded Lessons

- Value of early establishment of Silver level of emergency control in CPS.

- Need for separate managers located at physically divided work, viz. traffic management above and the propping work below the viaduct decks.

- Value of shielding the propping project team from the press, etc. to concentrate on their tasks.

- Value of recording all detail in pocket notebooks, provided the information is used rapidly.

- Set up clear management system and lines of communication as early as possible.

- Limit the number of managers in conference calls.

- Need for overall planning of the re-opening of the motorway carriageways.

Sources

M1 is fully reopened after Mill Hill scrapyard fire, BBC News, http://www.bbc.co.uk/news/uk-england-london-13154063, accessed 4 September 2011.

Scrapyard blaze forces M1 bridge closure. 2011. *New Civil Engineer*, 21 April.

Acknowledgements

Thanks are due to Connect Plus Services for information and advice on this project.

Appendix 1.7

Embankment Stabilization, Heck, Yorkshire, 2000–2001

Scope

The project consisted of urgent emergency work to stabilize a 1.8 km stretch of railway embankment carrying the UK East Coast Main Line which had been destabilized by up to 2 m depth of flood water washing against the foot of the embankment.

- Arrange access at both ends of the embankment.

- Create wider berms along each side of the embankment.

- Commence rock fill along the berms.

- Bulk fill of 200,000 tonnes.

- 450 mini-piles at bridges.

- Relocation of power cables, land drainage and a badger sett.

- New fencing, cleaning up.

Triggering Event

Following unprecedented heavy rainfall over much of Yorkshire in November 2000, the River Aire at Gowdall, North Yorkshire, overflowed its banks and filled the emergency washlands. One washland embankment then failed, flooded the village of Gowdall and created a flood plain equal in area to Lake Windermere (England's largest natural lake). It surrounded a 1.8 km length of the East Coast

Main Line at Heck. As a result, the railway embankment was destabilized by up to 2 m depth of water washing against the foot of the embankment. About 60 % of the embankment side slopes had slipped by up to 150 mm, cable routes had been disturbed and the supports to one underbridge disturbed.

Timetable

10 November 2000	Cracks and subsidence of embankment reported. Railtrack Zone Operations Department asks on-call consulting engineers to assess the embankment and asks the Project Department to manage the engineering work which will be needed.
11 November	Project staff and consultants reconnoitre site by helicopter. Contractor appointed to undertake site investigation, design and construction. Action plan agreed by all the above parties. Consultant appointed to manage safe access and supervise the contractor.
13 November	Access arranged with landowners. Full site investigation started.
20 November	Line re-opened for restricted speed working.
January 2001	Train speeds increased.
26 February	Full line speed restored on Up ECML.
12 March	Full line speed restored on Down ECML.

Leading Stakeholders

Railtrack plc was owner and controller of the track. They had overall responsibility for maintenance of the track, power and signalling systems, ensuring they could be used safely, and for emergency measures such as speed restrictions if required.

The train operating companies (TOCs) as users were the other major stakeholders in the line, but TOCs do not own the track. They are linked to Railtrack by contract.

Railtrack and TOCs are subject to the safety requirements of the Railway Inspectorate.

Project Organization

Within Railtrack the area was the responsibility of the ECML Zone Director. An immediate decision was made in Railtrack that management of the work would be the responsibility of the Railtrack Projects Department based in York. That department was responsible for all major projects for the East Coast Main Line (ECML).

Because of the urgency of the work a dedicated Project Manager was appointed, directly responsible to the Head of the Projects Department. The Project Manager was the senior engineer who was responsible for bridge and other civil engineering work in the area.

Railtrack Property Department arranged emergency access, planning permission and agreement with landowners.

Resourcing

A contractor who had just completed a rail bridge reconstruction project nearby satisfactorily was appointed to undertake site investigation, design and construction to stabilize the embankment and other affected structures. For site investigation and design the contractor employed consulting engineers who had considerable experience of similar railway work.

The contractor began work on the embankment under a variation order to the existing contract for the nearby bridge reconstruction work. Following this start, a contract for the Heck work was made under a 'rapid response' version of the Institution of Chemical Engineers 'Green Book' model conditions for reimbursable payment contracts, as used by Railtrack for unexpected work. The contractor was responsible for design and construction but subject to approval of proposals and method statements by Railtrack and the Railway Inspectorate.

A separate consultant also with extensive railway expertise was appointed to check design, supervise construction and monitor safety on behalf of Railtrack.

To establish good relationships rapidly it was agreed with the contractor and the consultants that they would bring known senior staff to the project.

Performance

The flooding at Heck occurred on a Friday. The ECML was already closed because of flooding further south, but that was expected to be cleared after the weekend. Restoration of use of the line at Heck was particularly urgent in order for the train operating companies to re-establish their services after restrictions imposed following the Hatfield derailment and by the widespread flooding.

The project team and contractor were brought together within 18 hours of the first call for their services. Some rock fill to start to consolidate the berms on each side of the embankment could be started before geotechnical investigation, analysis and design had been completed. For the main work a very large volume of rock and topsoil was required. As other means of access were flooded, most of the fill could delivered only by the local road through Henshall village. The community was already suffering from the flooding in the whole surrounding area. It was therefore agreed with the local authority that working would be limited to 14 hours per day, despite the urgency of restoring the ECML. At the peak, 250 lorry loads of rock and topsoil were brought through Henshall village daily. Over the three months the site was closed for only five days over Christmas and the New Year.

The line was re-opened five working days after the initial helicopter reconnaissance. The embankment, track and cabling were fully restored and 125 mph full-speed working resumed in three and a half months.

All the work was completed within budget.

There were no reportable accidents.

Sources

Environment Agency, North East Region. 2001. *Autumn 2000 Floods Review.* Regional Report.

Railtrack. 2001. *Heck Embankment Stabilisation Project.* East Coast Main Line, submission to National Rail Awards.

Railtrack. 2001. *Project Delivery.* Review.

Appendix 1.8

Remote Bridge Repair, Arnhem Highway, Northern Australia, 1998

Scope

To make and install a temporary deck structure spanning the ruptured central spans of a remote major highway bridge over a crocodile-infested tidal river. Urgent action pending pier repair and permanent replacement spans, plus some upgrading of other bridge piers. The location was remote from habitation or engineering support facilities.

Triggering Event

Underwater corrosion of some steel tubular piers supporting the bridge had caused partial collapse of these piers and rupture of bridge spans. The steel had corroded seven times faster than expected in these underwater conditions. The bridge was completed in 1972. It was the only all-weather land link between the mining, military and tourist region of North-East Arnhemland and Darwin. The collapse occurred at the peak season for traffic.

Timetable

The work to construct the temporary deck structure was effectively in three stages:

1.   Assessment of the bridge and decision on temporary replacement of spans.

2.   Preparation of temporary girders and other fabrication work off site, driving of new piles and strengthening of piers on site.

3.   Installation of girders, decking, etc.

Day 0	Department's engineers on site to assess the state of the bridge.
Day 1	The Department appointed the consulting engineers Sinclair Knight Merz as project manager.
Day 2	Experts' assessments and investigations began. Temporary pontoon passenger ferry provided.
Day 3	Choice of method of temporary bridge decided, based on using two 39 m long girders available from completed temporary work for a project in Darwin, on hire.
Day 5	Main design work completed.
Day 6	Contractors chosen. Long alternative track route opened for four-wheel-drive traffic.
Day 11	Temporary barge crossing service commenced.
Day 13	Pile driving started on damaged piers.
Day 31	Temporary bridge opened for all traffic except fully loaded road trains.
Day 43	Temporary bridge removed.
Day 44	Repaired bridge opened for limited day traffic.
Day 53	Load restrictions removed from repaired bridge.
Day 81	Bridge re-opened all hours to all traffic as a single-lane bridge. Barge and ferry services terminated.

Leading Stakeholder

The leading stakeholder was the Department of Works, Northern Territory.

There was no doubt then or since that the bridge had to be repaired or replaced. Building a new bridge was rejected for reasons of time and cost. The cost was met out of the Department's annual budget.

Project Organization

An integrated project team was formed drawn from the Department's and the consultant's staff. The consultants provided engineering resources and project management. The Department provided overall direction, technical staff, procurement services, contract management, cost monitoring and inspection staff. The consultants established a site office to manage the contracts and supervise the work. The Department provided site services.

Management meetings were held in Darwin, daily for the first two weeks and then weekly, under the Department's chief executive and senior executives, which were attended by the project manager, national experts, representatives of government, other Northern Territory departments and stakeholders.

The Department provided the media with a constant flow of information.

Except for some specialist advisers, all members of the project team were dedicated full time to the project until their roles were completed.

Resourcing

Three local contractors were invited to tender, each to undertake distinct packages of the fabrication and site work as none at the time had the resources for an immediate start to all of it. Meanwhile, the project manager initiated the procurement of major and long-lead time materials.

The contractors invited to tender were selected for their proven track records in the type of work required. Payment was on schedules of rates.

The contractors and all site staff worked around the clock.

The project manager operated as the Superintendent's Representative ('the Engineer') with powers to instruct the contractors, deal with claims and certify payments. A senior engineer of the Department's staff was the Superintendent for the contracts, to provide an 'impartial' overview. Site supervision was provided by the Department's technical staff, reporting through to the Superintendent's Representative.

Quality assurance was applied throughout the temporary and permanent work.

Performance

The project team stated that the critical decisions were:

- Immediate authorization to start the project.

- To proceed with a temporary structure spanning the collapsed pier using long girders on hire.

- Piles for a new pier could be driven from the temporary bridge, piers strengthened and the collapsed spans replaced.

- Ordering materials in advance of appointing contractors.

- Division of the work between contractors with the capacity.

- Department's inspection, cost monitoring and contract management staff located on site.

- Early establishment of communication systems for the project and for all stakeholders and media.

The project achieved all its objectives of restoring an all-weather route accommodated within the budget established early in the work.

There were no lost-time accidents during the project.

The project won the Institution of Engineers (Australia) Engineering Excellence Award for 1999.

Recorded Lessons

- Value of local knowledge of resources, particularly of the availability of the two girders.

- Speed by choosing a simple design solution.

- Speed by using known contractors for work appropriate to their immediate capacity.

Sources

Institution of Engineers Australia seminar. Darwin, 2001.

Sinclair Knight Merz and Northern Territory Government Department of Transport and Works. 1999. *Adelaide River Bridge Repairs*. Institution of Engineers Australia.

Acknowledgements

Thanks are due to the Northern Territory Government Department of Transport and Works and to Sinclair Knight Merz for information and further advice on this project.

Temporary Power Line, Auckland Central Business District, 1998

Scope

To design and construct a temporary emergency electricity 160 MVA 110 V line 9.8 km long and connecting work to restore the power supply to the Auckland Central Business District (CBD) in 17 days, February–March 1998. The emergency project was based upon erecting a temporary power transmission line utilizing a railway tunnel and running an open-wire transmission line alongside an operating railway line.

Triggering Event

Thermo-mechanical failure of one of two parallel buried oil-filled cables, causing overloading and failure of the second cable. Parallel circuits using older gas-filled cables were already out of service.

The risk of overheating failure of the oil-filled cables in their buried conditions was not expected. There had been no previous fault with these circuits. Previous to the failure the system owner had initiated a long-term project to improve the security of the power supply by constructing a new permanent line into the CBD.

Other power circuits into the Auckland CBD were of limited capacity. Actions to bring in portable generation systems and reduce the load could only mitigate the crisis temporarily.

Leading Stakeholders

The leading stakeholders were Mercury Energy, the electricity distribution and retail utility company supplying the southern half of the city, and Alstom New Zealand, a power systems contractor.

The cost of the temporary project was met by the utility.

Timetable

The utility's priorities after the cable failures were to mobilize their staff to arrange alternative power supplies, cable repair, optimize the use of the available power, investigate the failures and reinforce the rest of the system.

The emergency project to erect a temporary power transmission line was proposed by Alstom NZ on Day 2. The project to construct a temporary line proposed by a contractor was accepted by the utility as needed when the failed cables could not be repaired within an acceptable time. The utility issued a letter of intent to them for a contract for this work on Day 7. The project and some additional work for the utility was completed in 21 days.

Stages:

1.   Proposal to utility from contractor.

2.   Confirmation of route and detailed design.

3.   Construction and testing of the temporary system.

Project Organization

The contractor with subcontractors and consultants provided the complete team to design and construct the temporary power line and linking work, while the utility's staff were busy on other immediate actions to mitigate the failure of the cables.

The contractor appointed a dedicated Project Manager with responsibility for coordinating all work for the temporary project. The utility named a senior manager as 'the Engineer' for the contract, with representatives on site.

The utility's and contractor's managers met daily throughout the temporary project.

Resourcing

The utility appointed the contractor to be responsible for the design, supply, construction and commissioning of the temporary system. The utility took responsibility for government consents, public relations and site access.

The contractor subcontracted design and civil construction work, but retaining the work expected to be time critical (procurement, pole erection, conductor stringing, substation work and commissioning). The contractor appointed consulting engineers to provide the design team. The design team was relocated to the main contractor's offices.

The contractor was employed under reimbursable terms of payment based upon standard terms of contract, with incentives for completion time, safety and cost control.

Performance

Speed was achieved by:

- Commitment by the utility and contractor staff ahead of a letter of intent.

- Acceleration of the first programme for the work.

- Completion of conceptual design and agreement within three days.

- Planning to anticipate problems and remove constraints.

- Immediate actions to procure critical items, some at the expense of other projects.

- Use of available materials.

- Contractor's proposal to string an open power line in a tunnel and alongside an operating railway line.

- Cooperation of railway and city authorities.

- Commencing construction before detailed design completed.

- Use of statutory emergency powers for access to land.

- Commissioning team appointed early in the project.

The contractor first expected that six to eight weeks would be allowed for the work. On hearing that the failed cables could not be repaired and would have to be replaced, the contractor accelerated his programme, including getting insulators air-freighted from the USA. By this means, the project and some additional work for the utility could be completed in 21 days.

The project achieved all its objectives. It was completed within the 21 days promised in the accelerated programme.

Recorded Lessons

- Cooperation between government, commercial and all parties was vital.

- Locate all project team together, viz. design and management on site.

- Management teams need enlargement to cope with extended or 24-hour working.

- Hold team meetings in the evenings to minimize disruption to work.

- The value of detailed coordination of logistics and safety along the length of the line in a busy urban area.

- Use terms of contract appropriate to the objectives, viz. time.

- Extensive subcontracting enabled flexible use of resources.

- Cost escalated with acceleration of the work.

- Need to obtain rapid and detailed information on work status and problems.

- Performance of management teams could not have been sustained much beyond the three weeks.

Sources

Gatland, R. 1998. At the heart of the operation. *Auckland Lights Out from Failure to Recovery*. Conference, EA Technology, Chester.

Hunt, G. 1998. Auckland CBD transmission line project. *Auckland Lights Out from Failure to Recovery*. Conference, EA Technology, Chester, October, 84–94.

Ministry of Economic Development, New Zealand. 2001. *Inquiry into the Auckland Power Supply Failure*. Public summary, revised September 2001, available at http://www.med.govt.nz/inquiry/publicsummary.html, last accessed 15 July 2002.

Appendix 1.10

Aire Banks Repair, Gowdall, Yorkshire, 2000

Scope

The work consisted of the temporary raising of a section of the UK River Aire banks, emergency repairs and strengthening of flood-retaining embankments and construction of evacuation sluices and outfalls, consisting of:

- Earthworks to raise the level of washland cross banks to isolate the breached section.

- Temporary piling to raise the level of the river banks and to prevent further spilling into the breached section.

- Fill to top up other low washland banks.

- Piling to strengthen one of the cross banks.

- Construction of evacuation sluices to allow flood water in the washland to be drained back into the river.

The scope of the work increased as the design was developed to meet the requirements of strengthening cross banks, restoring the breached bank, adding a permanent evacuation sluice and removing sandbags.

Triggering Event

September 2000 saw twice the normal rainfall over south Yorkshire and heavy rain throughout October saturated much of the region. A further series of low pressure weather systems at the end of October caused heavy rain resulting in extensive flooding. Over 250 mm of rain fell in 12 days. Rivers in the region rose to the highest levels ever recorded, causing extensive flooding in some areas.

The lower Aire is tidal. Near Gowdall village, the heavy flood flow in the river combined with high tides caused the river to spill over one bank into washland, as intended, to be drained out again when the river level fell at low tides. The water in one washland then became a major risk to Gowdall as a result of a major slippage in the back face of a washland barrier bank. Repair of the bank was considered to be too dangerous. Work to establish a temporary flood defence was under way when the slippage turned into a breach of the washland bank. The defences successfully held up to 2 m depth of water, but leakage filled the fields towards Gowdall. Extensive sandbagging was carried out to protect the village. The situation worsened through 5–6 November due to the sheer volume of water. When it was clear that major flooding could not be prevented, most of the village was evacuated. One hundred and five properties were flooded (most of the village).

Leading Stakeholders

For the purposes of flood defence the Aire is classified as a main river. For main rivers the Environment Agency are responsible inter alia for flood warnings, for advising national and local government and other parties on actions needed for flood control and for implementing approved projects. The Agency had permissive powers to do work on assets. The Agency therefore undertook the planning and execution of the emergency work at Gowdall.

Operational 'Bronze' level of emergency control was established by the Police with the Fire Brigade, Environment Agency and East Yorkshire County Council.

The cost of the emergency work was met from Exchequer funding administered by the Ministry of Agriculture, Food and Fisheries.

Timetable

29 October–6 November	Three separate heavy rainfall peaks over the region.
29 October 2000	Flood warnings issued for whole immediate and wider area.
30 October	Special Operations Rooms were opened in the region to establish emergency warning systems and control of operations jointly with all other authorities.
30 October	River Aire flood water spills into washland (as planned).
2 November	Major slippage reported in washland bank.
2 November	Work to use railway embankment to contain floodwater started.
3 November	Leakage from washland starts to fill fields near Gowdall and sandbagging commenced to protect the village. Section of washland bank collapsed, causing extensive flooding of Gowdall.
4 November	Operational 'Bronze' level of emergency control established by the Police with the Fire Brigade, Environment Agency, Snaith District Council and East Yorkshire County Council.
4 November	Controlled breach made in another bank to alleviate flood threat to Gowdall.
6 November	Gowdall village evacuated.
6 November	Pumping out from breached washland commenced.
9 November	Environment Agency appoints its Combined Capital Works (Phase 2) (CCW2) team to help plan and manage the major emergency work where needed.
27 November	Scope and allocation of emergency work defined at CCW2 meeting.
29 November	Large pumps (see below) installed to pump out from breached washland.
9 December	River bank raising and cross bank strengthening completed. Low spots in other washland banks repaired, so that the protection standard was restored to the pre-flood conditions.
18 December	Evacuation of flood water from washland completed.
31 January 2001	Emergency escape sluices and new outfall structure completed.

The priorities at Gowdall were to isolate the breached washland section, pump out the water and prevent further flood spill from the river into the washland before the next high tides due on 12 December.

The filling of the other low washland banks effectively isolated the failed section of the flood defence infrastructure, allowing time for permanent repairs.

Project Organization

The Agency's Operations Department in the region includes resources for normal maintenance of the main river banks and flood control system. By October these resources were being employed round the clock on sandbagging and other emergency work.

The floods damaged river banks and flood defences in several locations in the region. The scale of work required was much greater than could be carried out by the river maintenance staff. The Agency therefore appointed its CCW2 design and construction team to assist with the major emergency work needed over the whole region, including at Gowdall.

The CCW2 team had just been appointed by the Agency under its National Capital Works Programme to design and construct a planned programme of river and flood defence improvements in the region, under the specific direction of the Agency. The CCW2 team consisted of a major contractor and engineering consultants. This established team provided an immediate resource to commence assessing the needs at Gowdall. The contractor and consultants then formed a small supplementary team for the emergency work for Gowdall (and then for Selby and Barlby).

Some of the work planned for the CCW2 contractor and consultant team was already affected by the wet weather and consequent floods and so was not delayed by giving priority to the emergency work.

The CCW2 team reported to the Agency's Project Manager, National Capital Programme Management Service, as for their planned programme of work.

For the emergency work, the Project Manager reported to the Agency's Regional Capital Works Manager and the Area Flood Defence Manager.

Resourcing

The CCW2 contractor and consultant team was responsible for design and construction. The contractor was employed under the New Engineering and Construction Contract. For the Gowdall emergency work payment was cost-plus, under Option E. The contractor employed the consultants under the same terms of contract.

Repair and culvert work was undertaken by the Agency's maintenance resources.

Appointment of the CCW2 team for the emergency work at Gowdall provided an established design and construction team to plan and manage the work. CCW2 planned all the work, and undertook the design and construction of the major work. The contractor also assisted in installing the mobile pumps.

Performance

Repair of the slipped section of the washland bank was considered to be too dangerous; instead, actions were taken to isolate the breached washland.

The pumps used to pump out from the breached washland were some of the world's largest mobile pumps, from Holland. They returned about 30 million tonnes of water back into the river.

Towards the completion of work on new sluices, access was restricted by a widespread Foot and Mouth outbreak.

Fill to restore the breached washland bank was completed later in 2001.

Normal procedures under the CDM Regulations were followed for all the work, but quickly. The Planning Supervisor was already in place for the CCW2 contract. A health and safety plan and method statements were compiled and approved.

The isolation of the breached section and prevention of further flood spill from the river into the washland were completed before the December high tides.

The emergency escape sluices and the remedial work to provide temporary protection to the village were completed ahead of schedule before Christmas.

There were no reportable accidents during the work.

The work was carried out during a period of many demands on Environment Agency staff, their consultants and contractors and all authorities over the whole area for flood control and many other measures for flood protection.

Recorded Lessons

The success of the emergency work critically depended on:

- Strong cooperation between the Environment Agency, local authority, police, fire and rescue services during and immediately following the flood event.

- Professional competence of all staff.

- Division of the whole flooding problem into subprojects for completion by teams.

- Use of the contract provisions pre-arranged for planned works.

- Competence and willingness of consultants and contractors.

- Regular written briefings to the villagers on proposals and progress and a daily multi-agency 'help caravan' in the village for residents to get advice and help build personal relationships.

Sources

Environment Agency. 2001. *Autumn 2000 Floods Review*. Regional Report, North East Region.

Harman, J. et al. 2002. Floods – A new approach. *Proceedings of the Institution of Civil Engineers – Civil Engineering*, 150, May, 2–59.

Home Office. 1997. *Dealing with Disasters*, 3rd edition.

Lane, G. et al. 2001. *36th Conference of River and Coastal Engineers*. Keele University, London: Department of the Environment, Food and Rural Affairs.

Reinstatement of Railway, Great Heck, Yorkshire, 2001

Scope

The project consisted of restoring the Up and the Down main line tracks, overhead power and signalling systems over 1 km of the UK East Coast Main Line (ECML). The damage extended to both rail tracks, the overhead power systems, ground cables and signalling, over a 1 km length of track and minor damage to a road overbridge and farming property. The cleaning up included removing coal and diesel fuel spilled from the freight train and the restoring of the areas used temporarily by the emergency services and others. Removal and replacement of a large amount of ballast and foundations containing spilled diesel was avoided by agreement that much of it could be allowed to drain naturally.

The two tracks and systems were all of modern standard at the time of the collision. Railtrack therefore decided that they should be restored to the previous specification, so minimizing the time needed to restore them.

Triggering Event

A Land Rover and loaded trailer had left a carriageway of the M62 Motorway and continued along a steep road embankment and subsequently down a railway embankment onto the ECML and into the path of an express passenger train travelling southbound to London. The impact with the Land Rover caused the train to derail. The momentum of the train carried it forward, staying substantially in line and upright, until it hit a set of points which further deflected the train into the path of a freight train travelling northbound on the other track. As a result of the second impact, the express train became virtually completely derailed and descended the embankment into an adjacent field to the south side of the bridge. The locomotive of the freight train also became derailed into the garden of a railside property north of that overbridge on the Down side. The time span from the Land Rover coming onto the line and the impact between the two trains was about one minute. Six passengers and four train staff died.

Timetable

28 February 2001	Road vehicle on the track leads to train derailment and the collision. Railtrack Projects Department, York, designated to manage the restoration work. The main contractor is designated by Railtrack which will plan and carry out the restoration work.
2 March	End of Police and Health and Safety Executive investigations permitted Railtrack access to start the removal of the wrecked trains. Contractor issued first draft method statement for the restoration of track, power and signalling systems.
5 March	Contractor issued final method statement for the work.
6 March	Removal of wreckage completed.
7 March, am	Track handed over for the restoration work. Subcontractor started mini-piling and foundations for OHL (overhead line) masts. Completion of temporary relay on Down line to permit access to start OHL masts. Completion of mast foundations, masts, OHL wiring and registration. Completion of relaying Up line. Completion of relaying Down line.
11 March, pm	Track returned to operation.
14 March	Completion of minor final work (by night possessions).
18 March	Full line speed working resumed.

Leading Stakeholders

Railtrack plc was the owner and controller of the track. It had overall responsibility for maintenance of the track, power and signalling systems and for emergency measures such as speed restrictions if required.

The train operating companies (TOCs) and freight operating companies (FOCs) as users were the other major stakeholders in the line, but do not own the track. They are linked to Railtrack by track access agreement contracts.

One farm suffered direct damage, and all the Great Heck village area was disrupted during the emergency of the collision and the line restoration work.

It was expected that all the restoration costs would be met by the road vehicle driver's insurance.

Project Organization

Within Railtrack, the operation and maintenance of the track was the responsibility of the LNE Zone Director. On notification of the collision, Railtrack Emergency Services Department established the incident as requiring the 'Gold' level of emergency control. A Railtrack senior manager was designated as Incident Manager to liaise on all decisions on clearing and restoring the line.

Following the experience of the Hatfield derailment, an immediate decision was made on the 28 February by Railtrack Zone management that direction of the Great Heck restoration work would be the responsibility of the Railtrack Projects Delivery Department based in York. That department was responsible for all major projects for the East Coast Main Line (ECML) and had taken over the management of the restoration work at Hatfield after the derailment there in October 2000.

Because of the size and importance of this work the Head of the Projects Delivery Department initially took responsibility as the Project Manager. When a clear way forward was established this was delegated to another Project Manager in the project delivery organization.

The Railtrack project team consisted of an integrated team of project managers, engineers and cost engineers to provide 24-hour/7-day coverage. The team included staff responsible for major rail embankment emergency stabilization work nearby and could immediately start on the work needed for rebuilding the stanchion bases for the overhead lines.

A temporary project HQ office for Railtrack and the contractor's staff was established on site. Temporary access was arranged with landowners by Railtrack's Property Department. Railtrack's QSs recorded all the work done and labour, plant and materials used by the contractor and subcontractors.

Procurement

Railtrack employed a maintenance contractor for this section of track to provide inspection, maintenance, renewal and replacement as required of the track, power and signalling. Under such contracts the maintenance contractor is responsible for maintaining the railway to the required standard.

This event was not due to a track fault. Following their experience gained at Hatfield, the Projects Department decided to appoint another contractor as the main contractor to plan and execute the reinstatement work. This contractor had already been selected to take over the maintenance contract for the Great Heck zone of the ECML after the imminent expiry of the existing contract.

The contractor was responsible for assessing the extent of track replacement and repair needed, and for checking that the completed work met the standards. This total responsibility followed practice for maintenance contracts.

The contractor was instructed to give this work priority over other contracts with Railtrack.

Railtrack nominated a specialist contractor as subcontractor for mini-piling and foundations for mast bases, following their work on the nearby rail embankment stabilization.

Following experience from Hatfield, the contractor selected for the restoration work immediately appointed an experienced Construction Manager to the site to plan all the work. During the rescue stage of work the Construction Manager provided personnel support services and other assistance to the clearing-up work and forensic investigations before the removal of the trains. This made it possible to observe the extent of damage and to start to plan the restoration work needed and draft a proposed method statement.

On this basis the contractor issued a first draft programme and detailed method statement before taking over the site. Planning was on the basis of working three shifts throughout. Much of the restoration work after clearance of wreckage was normal for track improvement and maintenance. The critical path was the OHL work, starting with removing damaged masts and foundations and replacing them. The crucial resource was the rail relaying manpower and track layer train. The Up track was the less damaged, so the

Down track was repaired temporarily, and then in turn the Up and the Down tracks re-laid.

The line at this point consists only of the two main line tracks. Once handed over after having been cleared of the train wreckage, the section could therefore be used exclusively for the restoration work, without limitations for passing trains. But the corollary was that there were no other undamaged tracks available for equipment to work alongside the damaged track. Most equipment and materials for the work were brought in along the two tracks. Road access was available from a minor road and needed police control to keep it clear of sightseers.

The programme in the contractor's method statement showed the sequence of work for track relaying in detail. Overall durations were shown for the overhead line and signalling work as these were the responsibilities of separate departments in the contractor's organization.

The contractor's first method statement and programme indicated that the date for completing all the work would be the 13 March. On taking over the site the completion date was advanced by two days as a result of discussions between client and contractor, on the basis that work not essential for the safe resumption of traffic could be completed during night possessions. Until this remaining work was completed the restored main line was subject to speed restrictions.

The Project Manager attended daily meetings at site with the contractor's Director responsible for the work.

The contractor's Construction Manager was in charge on site, with responsibility for planning the mobilization and use of resources, particularly critical resources such as the deployment of the track renewal train. Three experienced construction managers were appointed to this role to cover the three shifts. The contractor's Operations Managers were responsible for managing the use of the resources in each shift.

The terms of payment were cost reimbursable. The contractor had to check in detail many claims for payment from minor suppliers.

Other contractors were employed for the removal of wreckage, the spilled coal and damaged trees around the wrecked trains.

Performance

The ECML was restored to restricted service within seven days of access to the site, and full line speed working seven days later.

Railtrack and the TOCs are subject to the safety requirements of the Railway Inspectorate. Though urgent, the restoration work was subject to all track work safety requirements. Much of the work consisted of standard operations for track renewal so that a risk assessment could be readily produced by the contractor. The contractor's method statement included requirements for site safety, activity analysis, hazard control, subcontractors' method statements and site communications.

The contractor was the Principal Contractor under the CDM Regulations. The Railtrack Project Manager acted as the Planning Supervisor.

The contractor made contingency arrangements for the risks of late possession, failure of critical plant, track not fit for temporary use, damage to lineside cables, lack of labour, additional resources if needed to recover the programme and demands for coordination and management.

There were no reported accidents on site during the restoration work.

Recorded Lessons

The client and contractor considered that the keys to the success of the project were:

- Clarity of responsibilities from the outset.

- Lessons from a previous project (Hatfield – see Appendix 2) known to client and contractor.

- Good cooperation with the emergency services and HM Railway Inspectorate.

- Both parties immediately appointed a strong team.

- Client Project Manager kept communications lines simple so that the contractor had few other parties to deal with.

- Appointment of a contractor with all responsibility for, and the necessary skills to complete, the full work scope.

- Team-building between all main parties.

- A realistic programme at the outset to take the pressure off highly stressed staff.

- Employment of one contractor facilitated integrated planning so that some OHL and signalling and telecommunications work could start during track work.

Sources

Health and Safety Executive. 2001. *Train Derailment at Great Heck near Selby, 28th February 2001, Interim Report.*

Railway Safety. January 2002. *Great Heck.* Report.

Acknowledgements

Thanks are due to Railtrack plc. and Jarvis Rail for information, advice and comments on drafts for this case summary.

Appendix 1.12

9/11 Pile Sift, Make Safe and Remove Operations, World Trade Center, New York, 2001–2002

Scope

To sift, make safe and remove 1.6 million tons of rubble, hazardous major structural elements and other wreckage, 'the pile', in order to search for survivors and remains and clear the 9/11 site, Manhattan, over 10½ months from September 2001.

The work was urgent initially to search for survivors and then to find all identifiable remains while clearing the site.

Triggering Event

Hijacked passenger aircraft were deliberately crashed into the New York World Trade Center twin towers causing extensive fires and then total collapse of both towers and also destruction or severe damage to many adjoining buildings and city services. The twin towers accommodated thousands of office workers.

Leading Stakeholders

The Mayor and City of New York became the leading stakeholder, supported by US Federal and NY State emergency funds and charitable payments.

When first faced with one tower on fire the NY Fire Department called the Deputy Commissioner (DC) of the City Department of Design and Construction requesting temporary construction of ground-level protection from falling debris, a normal call for their assistance.

Collapse of the two towers changed the emergency into a massive engineering-dominated operation to sift the still-burning pile of rubble, hazardous major structural elements and other wreckage to rescue survivors. Sifting had been begun by the surviving Fire crews, City Police and Port Police and their reinforcements, assisted by many skilled and unskilled volunteers, by hand with only limited portable equipment. They were not equipped to search and rescue in a deep compressed pile of structural and other debris. At this point, before formal instructions from the City authorities, the DC called in civil engineering contractors to assist them by bringing in heavy construction plant to lift and cut heavy structural elements, speed the removal of rubble, and clear much other potentially hazardous wreckage.

Timetable

In effect, the project evolved in three overlapping stages:

1. Search and rescue work by the emergency services and volunteers, by hand and portable tools.

2. Large-scale work led by four main contractors running intensive heavy-equipment-dependent operations in their areas.

3. One main contractor for the whole site through to completion, clean up and handover.

The change to dependence on heavy plant marked the transition from Stage 1 to Stage 2. Deploying resources site wide marked the transition from Stage 2 to Stage 3.

Project Organization

Management of search and rescue in fires is normally the role of the NY Fire Department. They and the Police organized the initial sifting using their portable equipment.

A transition of leadership and control took place as the work became dependent on heavy construction plant. Initially on site the DC arranged for the provision of engineering plant and expertise as a service to the Fire crews and Police. By Day 3, the DC was taking the lead in the decisions on the use of all resources, through twice-daily meetings where all parties met to state needs and problems. The DC had the understanding of the city administrative systems to procure the resources needed on site and the engineering and construction experience, including previous emergency work on a stadium, to lead the critical decisions on how to employ them. He became accepted as the leader on site. He established trust so that decisions were made and acted upon with minutes or memos. Separate meetings were arranged on safety rules for the work. Many specialist submeetings were arranged ad hoc, but centered on the DC's meetings. Throughout, the DC had to see that construction productivity was secondary to search, rescue and body recovery. The City's Emergency Operations Center had been located in one tower and therefore had also been destroyed, but its staff organized the provision of site support services.

The project team was entirely temporary and developed flexibly through the three stages of the work. The site management team formed around the DC, initially on the basis of his expertise in procuring the engineering manpower and plant needed on site and then his leadership in the decisions on how to employ them. Two-tier leadership evolved; the Commissioner for Design and Construction, the DC's chief, managed the off-site relationships with the Mayor, other City authorities, the media and all other stakeholders. The DC drew on the advice of engineering experts, contractors' top managers and emergency and safety specialists, most initially on a voluntary basis, to assess and advise all parties on the needs on site and neighbouring structures.

The scale and nature of the work were beyond the experience, expectations or training of all parties. The twice-daily meetings were the flexible and rapid means of understanding the scale of the disaster, learning what was needed, establishing the team, discussing problems, agreeing actions and keeping all informed.

Resourcing

Four contractors were chosen for their experience and resources in the city. One had been employed by the City Department of Design and Construction on a previous emergency repair of a football stadium. Their first plant was on site on Day 1.

Three of the contractors were employed as management contractors and one as main contractor, each employing subcontractors. On Day 3 the DC agreed with the contractors to divide the site into four operating quadrants, one per contractor. The contractors and many subcontractors rotated teams week by week and worked 2 × 12-hour shifts throughout, with systematic shift handover meetings.

The contractors were employed on a form of contract adapted from one used for the previous emergency repair of a stadium. The payment terms were reimbursable for 'time and materials' costs, depending on complete records of time and cost, plus a 2.75 % fixed fee. These contracts were agreed on Day 3. An emergency payment system was arranged for the City to pay the contractors within 48 hours of the DC's certification of invoices.

Many other contractors and individual construction workers came to the site voluntarily. Some were retained as expert advisers, for instance on heavy demolition. Most were asked to leave because of excess numbers, non-required or inappropriate skills.

Voluntary organizations and many skilled and unskilled volunteers provided edge-of-site and off-site support. A muster point for volunteers was established at the site boundary.

The main contractors' expertise in using heavy plant was critical not only for the clearing operations but also for emergency earth-filling along one boundary to prevent inward collapse of the surrounding ground and flooding into the 'bathtub' hole as it was emptied of rubble.

After two months one contractor left the site having completed his quadrant, the least complex. By that time routines had been evolved for the bulk of the work and one of the three remaining contractors was selected to take overall responsibility for managing the use of resources over the whole site.

The project was monitored by the Federal Emergency Management Agency, advised by the US Corps of Engineers. They and the City resisted political pressures to bring in a large West Coast contractor over those already making good progress with the work.

All resources returned to their previous employment, but the event led to major changes in the roles of Federal security agencies.

Performance

The project started from an unspoken acceptance of an enforced escalation of normal support of search-and-rescue operations. All parties agreed tacitly on sifting and removing the pile, initially as a search-and-rescue operation and then to sustain this to try to identify remains. The project met these objectives. Reviews since have not questioned the assumption that search and rescue should have priority, continuing through the months of work to sift for identifiable remains.

The project organization was entirely temporary and developed flexibly through the three stages of the work.

CRITICAL DECISIONS

- The Mayor and advisors supported the lead taken by the City Department of Design and Construction.

- The DC first estimated that the project would take a year and cost $1 billion. This was a shock to the City authorities, but when accepted provided the cover for the quick calls for resources needed to meet the emerging demands of the search and rescue work.

- Employing known, capable contractors.

- Agreement on reimbursable contract terms of payment.

- Close relationships with the media.

- Exceptional safety procedures agreed for certain conditions, temporarily, with employees and all parties including enforcement agencies.

Recorded Lessons

- Emergency plans are vital for identifying roles and resources, but not for prescriptions of what to do.

- Emergency communication links may be lost because systems are destroyed or overloaded.

- Bring in the right expertise.

- All should take regular breaks. The initial responses of cooperation and high level of performance were sustained over a long period, but with subsequent anxieties about physical and mental health.

- Unskilled resources are of little value on site but should be found off-site supporting work.

- Too many of the right resources can hamper progress and safety.

- Coordination is needed with operations beyond the site, viz. transport priorities, disposal routes and dumps, fuel supplies.

- Trust in accepting oral decisions is essential to speedy implementation.

- Communicate with all stakeholders.

- The value of the immediate response of trusted contractors and vendors.

- In reimbursable terms of payment every contractor and supplier should keep their own detailed records, from the start.

- Large immediate payments must be arranged.

- Site boundaries need to controlled, but in communication with the project manager to ensure entry of required personnel and resources.

- Access security control badges need to be changed regularly.

Sources

Federal Emergency Management Agency. 2002. *Critical Infrastructure*. Conference, Austin, Texas, February.

Fleming, D. 2002. Subway trains arrive early (New York World Trade Center). *New Civil Engineer*, 5, September, 24–5.

Langewiesche, W. 2002. *American Ground: Unbuilding the World Trade Center*. Farrar, Strauss & Giroux; published in UK by Scribner, 2003.

Major Projects Association. 2003. *The Management of Unexpected Projects*, seminar, London, June, summary published at http://www.majorprojects.org/pdf/seminarsummaries/108Unexpprojects.pdf, last accessed 25 June 2014.

National Committee on Terrorist Attacks. 2004. *The 9/11 Commission Report*. US Government Printing Office.

Rubin, D.K. 2002. Michael Burton – Engineer and award winner, citation, *Engineering News Record*. Available at http://www.engology.com/eng5burton.htm, last accessed November 2004.

Smith, D. 2002. *Report from Ground Zero*. New York: Doubleday.

Stewart, J.B. 2002. *Heart of a Soldier*. New York: Simon & Schuster [biography of Rick Rescorla, safety officer, Morgan Stanley, World Trade Center].

Appendix 2

References to Publications on Other Cases

Animal Epidemic Support Work, 1969

SCOPE

Deployment and accountability in diverting river maintenance and small works resources to aid other authorities in managing an animal epidemic.

SOURCE

Doody, M.C. 1968. *The Organization of Men and Machines for Emergencies*. MSc thesis, University of Manchester Institute of Science and Technology.

LESSON

Offer to divert resources frozen by an emergency to help recovery from that emergency may be delayed by risks of accountability.

Carlisle City Centre Flooding Recovery, 2005

SCOPE

Recovery, cleaning up and achieving return to business as normal hampered because city critical infrastructure, communications and emergency services were affected by convergent river flooding.

SOURCES

Carlisle City Council et al. (n.d.). *Carlisle Renaissance*. Prospectus.

Government Office for the North West. 2005. *Carlisle Storms and Associated Flooding*. Multi-agency debrief report.

Robertson, D. and Koronka, P. 2005. *The Great Flood – Cumbria 2005*. South Stainmore: Hayloft Publishing.

LESSONS

Emergency coordination may be stymied by the emergency. Restoration of communications and power is a priority. Need full recovery plan as part of business continuity and emergency planning. Need delegated authorities spend. Role of elected representatives and officers needs to be clear. Need systematic link with insurers. Value of a renaissance plan.

Daly River Infrastructure Recovery, Australia, 1998

SCOPE

Restoration of power, water, road and air services to bush communities by combined work of regional and other government bodies: defence forces, utilities and local business already stretched by extensive cyclonic flooding.

SOURCE

Rolland, D. 1998. *Infrastructure Recovery after the 1998 Daly River Floods*. National Engineering Award, Institution of Engineers Australia.

LESSONS

Embrace all stakeholders. Value of establishing a unified information secretariat. Defence Forces' resources have limited capacity.

Darwin Power Restoration, 1974

SCOPE

Restoration of electricity supply system in most of Darwin destroyed or severely damaged by Cyclone Tracey, drawing on teams from 19 Australian power authorities.

SOURCE

Peake, O. 1976. *Restoration of Electricity Supply Following Cyclone Tracey.* Institution of Engineers Australia.

LESSONS

Initially daily team briefings were necessary. Electricity system control centre wrecked. Resources brought in from different power authorities followed different procedures. Provision of food and transport for visiting crews consumed surprisingly large manpower. Local people provided valuable information and support. Restoration of the centre was a mix of temporary and permanent work.

East Timor Power Infrastructure Reinstatement, 1999

SCOPE

Restoration of power stations and power distribution system after systematic sabotage by departing foreign occupying forces.

SOURCE

Pemberton, C. 2000. *Reconstruction of the Power System in East Timor.* 76th annual conference, Electric Energy Society of Australia, Canberra.

LESSON

Interdependence of power, water and other wrecked infrastructure services requires integrated assessment and planning.

Hatfield Rail Track Reinstatement, 2000

SCOPE

Renewing and upgrading two main line and two slow line tracks, overhead power and signalling systems after derailment of InterCity express train caused by fractured lengths of rail.

SOURCES

Health and Safety Executive. 2001. *Train Derailment at Hatfield*. Second Interim Report.

Hibbert, L. 2002. Down the line from Hatfield. *Professional Engineering*, 13 February 2002, 28–30.

LESSONS

Value of the immediate appointment of an incident manager, to coordinate sponsor's and others' actions, for clearing wreckage, service substitutions and recovery. Recognition of managing as a project helped fast start to reinstatement work. Value in tight space of contractor who needed only limited use of others' resources.

Heathrow Express Rail Tunnels Recovery, 1994

SCOPE

Removal of collapsed partially complete shallow tunnel bores and unstable ground under London Heathrow Airport, reconstruction with changes of design, changes in construction method and switching New Engineering Contract responsibilities options.

SOURCES

Deane, A.P., Myers, A.G. and Tipper, G.C. 1997. *Heathrow Express – New Rail Tunnels under London's Main Airport*. Rapid Excavation and Tunnelling Conference, Las Vegas.

Finch, A.P. 1997. *Heathrow Express Recovery from Adversity*. Going Underground, Conference, Institution of Civil Engineers, Midlands Association.

Lownds, A. 1998. Organisational transformation on the Heathrow express, quoted in Winch, G.M. 2002. *Managing Construction Projects*. Oxford: Blackwell Publishing.

McBeth, D. 2002. Heathrow tunnel collapse – Seven years on. *Proceedings of the Institution of Civil Engineers – Civil Engineering*, August, 1449(3), 101.

LESSONS

Integrated team of sponsor, contractor, designers and insurers, with consultant hired to create a culture of trust. Value of alternative terms of payment under unchanged generic terms of contract plus single team agreement. Motive of all stakeholders to proceed urgently. Reconstruction with design changes to improve safety and speed.

Kobe (Great Hanshin-Awaji) Earthquake Recovery, 1995

SCOPE

Rebuilding industrial areas, port and communities after an earthquake in an area in which seismic hazards had not been a major focus of governmental or public concern. Large-scale volunteer activity in restoring communities and employment was not anticipated in government plans.

SOURCES

Hashimoto, N. 2000. Public organizations in an emergency: The 1995 Hanshin-Awaji earthquake and municipal government. *Journal of Contingencies and Crisis Management*, 8(1), 15–22.

Shaw, R. and Goda, K. 2004. From disaster to sustainable civil society: The Kobe experience. *Disasters*, 28(1), 16–40.

Tierney, K.J. and Goltz, J.D. 1997. *Emergency Response: Lessons Learned from the Kobe Earthquake*. Disaster Research Center, University of Delaware.

LESSONS

Need to inform and involve all stakeholders. Priority given to business over social needs. Culture change towards recognizing value of local community action.

Manchester Bomb Recovery Work, 1997

SCOPE

Making safe and strengthening shopping centre structure after large terrorism bomb.

SOURCES

Roditakis, E. 1997. *Arndale Recovery Project*. MSc dissertation, University of Manchester Institute of Science and Technology.

Sandford, A. 1997–1998. Bombed not bowed. *Memoirs of the Manchester Literary and Philosophical Society*, 136, 51–9.

LESSONS

Rapid response to create stakeholder collective. Value of management contractor in place with up-to-date knowledge of main structure. Stimulus to establishing a city centre renaissance plan and project management competence.

Motorway Viaduct Emergency Repair, 2012

SCOPE

Repair of 100+ cracks in the 1 km-long massive electro-slag welded steel M4 motorway overhead viaduct connecting Heathrow Airport and central London in three months before the London Olympics.

SOURCE

Wearne, C.P. et al. 2013. Collaboration in an Emergency: Repairing London's M4 Boston Manor Viaduct. *Proceedings of the Institution of Civil Engineers – Civil Engineering*, 166(CE1), 33–9, February.

LESSONS

Immediate formation of unified team of all stakeholders. Benefit of Lean technique with facilitator leading daily collaborative detailed planning.

Value of established framework contracts. Daily meetings to aid integration across organisational boundaries.

New York Subway Station and Track Restoration, 1991

SCOPE

Make safe, clean up and rebuild supporting columns, track, power and signal systems after destruction by derailment of five cars of train in subway tunnel.

SOURCES

Meredith, J.R. and Mantell, S.J. 1995. *Project Management – A Managerial Approach*. 3rd edition, Hoboken, NJ: John Wiley.

Nacco, S. 1992. PM in crisis management at NYCTA: Recovering from a major subway accident. *PMNETwork*, February 1992, 9–27.

LESSONS

Command and control immediately established around role of Wreckmaster. Person appointed had experience of multi-stakeholder reconstruction project. Value of work breakdown principle in minds of managers as basis of delegation of control. Test completed repairs under expert eyes.

Northridge Utilities Services Restoration, Los Angeles, 1994

SCOPE

Restoration of freeways, water, power distribution and many utilities and local related services after 'the most devastating seismic event in Los Angeles history'.

SOURCES

Baxter, J.B. 1994. Responding to the North Ridge earthquake. *PMNETwork*, November, 13–22.

Ranous, R.A. 1995. Post-earthquake safety assessment: Deploying qualified personnel following the Northridge earthquake. *Building Standards*, 8–12.

Schiff, A.J. (ed.). 1995. *Northridge Earthquake – Lifeline Performance and Post-Earthquake Response*. American Society of Civil Engineers.

LESSONS

Managerial task force coordinated all agencies. No agency rivalry. Rough estimate of damage repair costs overwhelmingly less than service revenue loss. Temporary substitute highway services permanently affect long-term demand. Poor inventory of spares. Overall program manager appointed. State Governor overrode 'the usual lengthy contract process for construction'. Force Account contracts for demolition. Informal bid cost + time bonus/penalty contracts for reconstruction. Pre-selected only experienced contractors who agreed to work on only one contract. Bids within hours or days. Contractors responded to contract time incentive.

Thunder Horse Drilling Riser Replacement, Gulf of Mexico, 2003

SCOPE

Respond to the threat to people and environment, secure the remaining riser section, recover pieces and reconnect rig to well-head at a depth of 3,200 ft.

SOURCE

Crichton, M.T., Lauche, K. and Flin, R. 2005. Incident command skills in the management of an oil industry drilling incident: A case study. *Journal of Contingencies and Crisis Management*, 13(3), 116–28.

LESSONS

Value of control directed by Incident Management Team. Subteams for each work package. Incident control on-shore. Strategic decisions made by Incident Control Manager, tactical by deputy and subteam leaders, and operational decisions made on the rig. Leadership of meetings critical. Need for domain-specific knowledge and incident command skills. Team members should be given performance feedback.

Warship Recommissioning, 1982

SCOPE

Acceleration of ship refit and recommissioning for special large role.

SOURCE

Larken, E.S.J. 2002. Military commander – Royal Navy, in Flin, R. and Arbuthnot, K., *Incident Command: Tales from the Hot Seat*. Farnham: Ashgate Publishing.

LESSONS

Ensure engineering complete under time pressure. Need to plan preparation to be leader of newly formed much larger mixed team and to provide support for top leadership.

Webbers Falls Bridge Replacement, I-40, Oklahoma, 2002

SCOPE

Replacement of end piers and spans of critical interstate highway bridge downed by an errant barge-pusher boat.

SOURCES

Bai, Y., Burkett, W.R. and Nash, P.T. 2006. Lessons learned from an emergency bridge replacement project. *Journal of Construction Engineering and Management*, 132(4), 338–44.

Bai, Y., Kim, S.H. and Burkett, W.R. 2007. Enhancing the capability of rapid bridge replacement after extreme events. *Engineering, Construction and Architectural Management*, 14(4), 375–86.

Yang, D. 2003. *Rapid Bridge Replacement Techniques*. Region 4 Research Advisory Committee Meeting, American Association of State Highway and Transportation Officials, August 5, San Antonio, TX.

LESSONS

Value of system triage. Analyse when alternative routes are the priority over restoring lost facility. Design contract for bridge spans replacement placed on day of accident, fixed price with schedule incentive and penalty. Challenges included coordination of fast design work between steel detailer, sponsor and checkers. Replacement spans design different to originals to expedite the process. Used established procedures and innovative contracting. Reconstruction contract fixed price plus schedule bonus/penalty. Sponsor's retired personnel employed to augment inspection staff. Contractor paid safety bonus. Good commitment of all parties. Strong support from the local community. The lessons contributed to national strategies and technologies for the quick restoration of highway bridges.

References

Note: The sources used in each case are listed in that case summary.

APM (Association for Project Management) 2011. *Directing Change: A Guide to Governance of Project Management*, 2nd edition. Association for Project Management.

Atkins, S. and Gilbert, G. 2003. The role of induction and training in team effectiveness. *Project Management Journal*, 34(2), 48–52.

Bai, Y., Burkett, W.R. and Nash, P.T. 2006. Lessons learned from an emergency bridge replacement project. *Journal of Construction Engineering and Management*, 132(4), 338–44.

Barnes, N.M.L. 1971. Lecture, Project Management Group, University of Manchester Institute of Science and Technology; later published in *International Project Management Yearbook*, 1985. International Project Management Association.

Baxter, J.B. 1994. Responding to the North Ridge earthquake. *PMNETwork*, November, 13–22.

Beer, S. 1972. *The Brain of the Firm*. London: Allen Lane, The Penguin Press.

Bentley, C. 2009. *PRINCE2 Handbook*, 3rd edition. APMG Business Books.

Bonke, S. and Winch, G.M. 2000. A mapping approach to managing project stakeholders. *Proceedings of PMI Research Conference*. Paris, Project Management Institute.

Bower, D. (ed.). 2003. *Management of Procurement*. London: Thomas Telford Ltd.

British Standards Institution (BSI). 2011. *PAS 200:2011, Crisis Management: Guidance and Good Practice*.

Byfield, M. et al. 2008. Learning from failures. *Proceedings of the Institution of Civil Engineers*, 161, special issue.

Carter, W.N. 1991. *Disaster Management*. Asian Development Bank.

Cooper, R.G. 1993. *Winning at New Products: Accelerating the Process from Idea to Launch*. New York: Perseus Books.

Dalton, G.E. 2003. Private communication.

Dawkins, C.R. 1976. *The Selfish Gene*. Oxford: Oxford University Press.

Eastham, G.E. 2002. Private comment.

ECI (European Construction Institute). 2002. *Fast Track Manual*. European Construction Institute.

Eijkenaar, J.W.D. 1997. *Project Management in Complex Emergencies*. MSc dissertation, University of Manchester Institute of Science and Technology [management of humanitarian aid projects].

Ellis, B. 2012. *10 Rules of Crisis Management*. EVP/Crisis Communications and whatcanbe Lab, USA, cited on DIIT Forum, 19 September 2012, at: http://forum.diit.info/index.php/topic,3207.msg3496.html?PHPSESSID=tav8lgoqs spth2duaqt1ohdc80#msg3496, last accessed 7 April 2014.

Engwall, M. and Svensson, C. 2003. *Cheetah Teams: The Most Extreme Form of Temporary Organization*. European Academy of Management (EURAM), annual conference, April, Milan.

Flin, R. and Arbuthnot, K. 2002. *Incident Command: Tales from the Hot Seat*. Farnham: Ashgate Publishing.

Flyvbjerg, B. 2006. Five misunderstandings about case-study research. *Qualitative Inquiry*, 12(2), 219–45.

Flyvbjerg, B. 2011. Over budget, over time, over and over again: Managing major projects, in Morris, P.W.G., Pinto, J.K. and Söderlund, J. (eds), *The Oxford Handbook of Project Management*. Oxford: Oxford University Press, 321–44.

Gaddis, P.O. 1959. The project manager. *Harvard Business Review*, May–June, 89–97.

Galbraith, J. 1977. *Organization Design*. Boston, MA: Addison-Wesley.

Geraldi, J.G. et al. 2010. The Titanic sunk, so what? Project manager response to unexpected events. *International Journal of Project Management*, 28(6), 547–58.

Gladwell, M. 2008. *Outliers*. London: Penguin Books.

Handy, C. 1987. *Understanding Organizations*. London: Penguin Books.

Hashimoto, N. 2000. Public organizations in an emergency: The 1995 Hanshin-Awaji earthquake and municipal government. *Journal of Contingencies and Crisis Management*, 8(1), 15–22.

Howard, A. (ed.). 1979. *The Crossman Diaries*. London: Book Club Associates.

IPMA (International Project Management Association). 1987. *Handbook of Project Start-Up*, ed. Fangel, M. International Project Management Association.

Kim, J. and Burton, E.M. 2002. The effect of task uncertainty and decentralization on project team performance. *Computational and Mathematical Organization Theory*, 8(4), 365–84.

Kingdon, D.R. 1973. *Matrix Organization*. London: Tavistock Publications.

Knight, R.F. and Pretty, D. 1996. *The Impact of Catastrophes on Shareholder Value*. Report, Oxford Metrica, Templeton College, University of Oxford.

Langewiesche, W. 2002. *American Ground: Unbuilding the World Trade Center*. New York: Farrar, Strauss & Giroux; published in UK by Scribner, 2003.

Larken, E.S.J. 2002. Military commander – Royal Navy, in Flin, R. and Arbuthnot, K., *Incident Command: Tales from the Hot Seat*. Farnham: Ashgate Publishing.

Lawrence, P. and Lorsch, J. 1967. Differentiation and integration in complex organizations. *Administrative Science Quarterly*, 12, 1–30.

Lawrence, P.R. and Lorsch, J.W. 1967. *Organization and Environment*. Boston, MA: Harvard Business Press.

Le Masurier, J., Wilkinson, S. and Shestakova, Y. 2006. *An Analysis of the Alliancing Procurement Method for Reconstruction Following an Earthquake.* Proceedings of the 8th US National Conference on Earthquake Engineering, April 18–22, San Francisco.

Lewin, C. et al. 2006. Risk: Facing the reality. *Proceedings of the Institution of Civil Engineers,* 159, special issue 2.

Lienart, A. 1996. Coping with disaster – Federal Employees Credit Union. *Management Review,* 85(11), 38–42.

Loosemore, M. 1998. Organisational behaviour during a construction crisis. *International Journal of Project Management,* 16(2), 115–21.

Lovallo, D. and Kahneman, D. 2003. Delusions of success. *Harvard Business Review,* July, 56–63.

Mantel, S.J. 2003. Private communication on data reported in Meredith, J.R. and Mantel, S.J. 2000. *Project Management: A Managerial Approach,* 5th edition. Hoboken, NJ: John Wiley & Sons.

Margerison, C. and McCann, R. 1984. The managerial linker: A key to high performing teams. *Management Decision,* 22(4), 42–9.

McDonough, E.F. and Pearson, A.W. 1993. An investigation of the impact of perceived urgency on project performance. *Journal of High Technology Management Research,* 4(1), 111–21.

Melkonian, T. and Picq, T. 2010. Opening the 'Black Box' of collective competence in extreme projects: Lessons from the French Special Forces. *Project Management Journal,* 41(3), 79–90.

Morris, P.W.G. and Hough, G.H. 1987. *The Anatomy of Major Projects: A Study of the Reality of Project Management.* Chichester, New York: John Wiley & Sons.

NAO (National Audit Office). 2002. *Better Public Services through e-government: Academic Article in support of Better Public Services through e-government.* Report by the Comptroller and Auditor General, HC 704-III Session 2001–2002.

Nobelius, D. and Trygg, L. 2002. Stop chasing the front end process – management of the early phases in product development projects. *International Journal of Project Management*, 20(5), 331–40.

Office of Government Commerce (OGC). 2004. *NHS National Programme for IT (Strategic Assessment)* [public release of a series of 'Gateway Reviews' introduced in 2002].

Philips, P. 2005. *Lessons for Post-Katrina Reconstruction*. Briefing paper, Washington DC: Economic Policy Institute.

Pilcher, R. 1967. *Principles of Construction Management for Engineers and Managers*. New York: McGraw-Hill.

PMI (Project Management Institute). 2005. *Project Management Methodology for Post Disaster Reconstruction*. Project Management Institute.

Prieto, B. 2011. 'Black Swan' risks. *PM World Today*, XIII(1).

Rickards, T. and Clark, M. 2006. *Dilemmas of Leadership*. New York: Routledge.

Shenhar, A.J. 2001. One size does not fit all projects: Exploring classical contingency domains. *Management Science*, 47(3), 394–414.

Söderholm, A. 2008. Project management of unexpected events. *International Journal of Project Management*, 26, 80–86.

Solnit, R. 2009. *A Paradise Built in Hell – The Extraordinary Communities that Arise in Disaster*. New York: Viking.

Stretton, A.B. 1979. In *Natural Hazards in Australia*, Australian Academy of Science.

Terwiesch, C. and Loch, C.H. 1999. Managing the process of engineering change orders: The case of the climate control system in automobile development. *Journal of Product Innovation Management*, 16, 160–72.

Tuckman, B.W. 1965. Development sequence in small groups. *Psychological Bulletin*, 63(6), 384–99.

Turner, J.R. 2006. Towards a theory of project management: The nature of the project governance and project management. *International Journal of Project Management*, 24(2), 93–5.

Turner, J.R. and Cochrane, R.A. 1993. Goals-and-methods matrix: Coping with projects with ill-defined goals and/or methods of achieving them. *International Journal of Project Management*, 11(2), 93–102.

Turner, J.R. and Müller, R. 2003. On the nature of the project as a temporary organization. *International Journal of Project Management*, 21(1), 1–8.

Ward, S. and Chapman, S. 1994. Choosing contractor payment terms. *International Journal of Project Management*, 12(4), 216–21.

Warren, B. and Biederman, P.W. 1997. *Organizing Genius: The Secrets of Creative Collaboration*. New York: Perseus Books.

Wearne, S.H. 1970. Principles in organizing design staff. *Proceedings of the Institution of Mechanical Engineers*, 180(3M), 72–80.

Wearne, S.H. 1993. *Principles of Engineering Organization*, 2nd edition. London: Thomas Telford Ltd.

Wearne, S.H. 2002. Management of urgent emergency engineering projects. *Proceedings of the Institution of Civil Engineers – Municipal Engineer*, 151(4), December, 255–63. Reprinted in *IEEE Engineering Management Review*, 2005, 33(3), 21–31.

Wearne, S.H. 2014. Evidence-based scope for reducing 'fire-fighting' in project management. *Project Management Journal*, 45(1), 67–75.

Weick, K.E. 1993. The collapse of sense-making in organizations: The Mann Gulch disaster. *Administrative Science Quarterly*, 38(4), 628–52.

Weick, K.E. and Sutcliffe, K.M. 2001. *Managing the Unexpected: Resilient Performance*. Hoboken, NJ: John Wiley & Son.

Wijngaard, P., Mooi, H. and Scholten, V. 2010. Project managers and executives. *Project Perspectives 2010*. International Project Management Association.

Williams, T. and Knut, S. 2010. Issues in front-end decision making on projects. *Project Management Journal*, 41(2), 38–49.

Winch, G.M. 2002. *Managing Construction Projects*. London: Blackwell Publishing.

Zhang, H. and Flynn, P. 2003. Cited in Eriksson, Effectiveness of alliances between operating companies and engineering companies. *Project Management Journal*, 34(3), 48–52.

Zwikael, O. and Unger-Aviram, E. 2010. HRM in project groups: The effect of project duration on team development effectiveness. *International Journal of Project Management*, 28(6), 413–21.

Index

Advances in Project Management

Advances in Project Management provides short, state of play, guides to the main aspects of the new emerging applications including: maturity models, agile projects, extreme projects, Six Sigma and projects, human factors and leadership in projects, project governance, value management, virtual teams and project benefits.

Currently Published Titles

Advances in Project Management, edited by Darren Dalcher 978-1-4724-2912-4

Project Ethics, Haukur Ingi Jonasson and Helgi Thor Ingason 978-1-4094-1096-6

Managing Project Uncertainty, David Cleden 978-0-566-08840-7

Managing Project Supply Chains, Ron Basu 978-1-4094-2515-1

Project-Oriented Leadership, Ralf Müller and J Rodney Turner 978-0-566-08923-7

Strategic Project Risk Appraisal and Management, Elaine Harris 978-0-566-08848-3

The Spirit of Project Management, Judi Neal and Alan Harpham 978-1-4094-0959-5

Sustainability in Project Management, Gilbert Silvius, Jasper van den Brink, Ron Schipper, Adri Köhler and Julia Planko 978-1-4094-3169-5

Second Order Project Management, Michael Cavanagh 978-1-4094-1094-2

Tame, Messy and Wicked Risk Leadership, David Hancock 978-0-566-09242-8

Reviews of the Series

PROJECT ETHICS, HAUKUR INGI JONASSON AND HELGI THOR INGASON

> *This book will instil more confidence in the PM profession and will help individuals become better practitioners.*
>
> PM World Journal, vol. II, no. V

> *Project leaders managing any type of project but especially large complex projects with a diverse stakeholder group would benefit from this book. By adding the ethical analysis to the risk assessment management plan, the project leader will consider the broader implication of the project.*
>
> PM World Journal, vol. III, no. VII

MANAGING PROJECT UNCERTAINTY, DAVID CLEDEN

> *This is a must-read book for anyone involved in project management. The author's carefully crafted work meets all my '4Cs' review criteria. The book is clear, cogent, concise and complete … it is a brave author who essays to write about managing project uncertainty in a text extending to only 117 pages (softcover version). In my opinion, David Cleden succeeds brilliantly…For project managers this book, far from being a short-lived stress anodyne, will provide a confidence-boosting tonic. Project uncertainty? Bring it on, I say!*
>
> International Journal of Managing Projects in Business

> *Uncertainty is an inevitable aspect of most projects, but even the most proficient project manager struggles to successfully contain it. Many projects overrun and consume more funds than were originally budgeted, often leading to unplanned expense and outright programme failure. David examines how uncertainty occurs and provides management strategies that the user can put to immediate use on their own project work. He also provides a series of pre-emptive uncertainty and risk avoidance strategies that should be the cornerstone of any planning exercise for all personnel involved in project work.*
>
> *I have been delivering both large and small projects and programmes in the public and private sector since 1989. I wish this book had been available when I began my career in project work. I strongly commend this book to all project professionals.*
>
> Lee Hendricks, Sales & Marketing Director,
> SunGard Public Sector

The book under review is an excellent presentation of a comprehensive set of explorations about uncertainty (its recognition) in the context of projects. It does a good job of all along reinforcing the difference between risk (known unknowns) management and managing uncertainty (unknown unknowns – 'bolt from the blue'). The author lucidly presents a variety of frameworks/ models so that the reader easily grasps the varied forms in which uncertainty presents itself in the context of projects.

VISION – The Journal of Business Perspective (India)

Cleden will leave you with a sound understanding about the traits, tendencies, timing and tenacity of uncertainty in projects. He is also adept at identifying certain methods that try to contain the uncertainty, and why some prove more successful than others. Those who expect risk management to be the be-all, end-all for uncertainty solutions will be in for a rude awakening.

Brad Egeland, *Project Management Tips*

PROJECT-ORIENTED LEADERSHIP, RODNEY TURNER AND RALF MÜLLER

Müller and Turner have compiled a terrific 'ready-reckoner' that all project managers would benefit from reading and reflecting upon to challenge their performance. The authors have condensed considerable experience and research from a wide variety of professional disciplines, to provide a robust digest that highlights the significance of leadership capabilities for effective delivery of project outcomes. One of the big advantages of this book is the richness of the content and the natural flow of their argument throughout such a short book…Good advice, well explained and backed up with a body of evidence…I will be recommending the book to colleagues who are in project leader and manager roles and to students who are considering these as part of their development or career path.

Arthur Shelley, RMIT University, Melbourne, Australia,
International Journal of Managing Projects in Business

In a remarkably succinct 89 pages, Müller and Turner review an astonishing depth of evidence, supported by their own (published) research which challenges many of the commonly held assumptions not only about project management, but about what makes for successful leaders.

This book is clearly written more for the project-manager type personality than for the natural leader. Concision, evidence and analysis are the main characteristics of the writing style…it is massively authoritative, and so carefully written that a couple of hours spent in its 89 pages may pay huge dividends compared to the more expansive, easy reading style of other management books.

Mike Turner, Director of Communications for NHS Warwickshire

STRATEGIC PROJECT RISK APPRAISAL AND MANAGEMENT, ELAINE HARRIS

… Elaine Harris's volume is timely. In a world of books by 'instant experts' it's pleasing to read something by someone who clearly knows their onions, and has a passion for the subject … In summary, this is a thorough and engaging book.

Chris Morgan, Head of Business Assurance
for Select Plant Hire, Quality World

As soon as I met Elaine I realised that we both shared a passion to better understand the inherent risk in any project, be that capital investment, expansion capital or expansion of assets. What is seldom analysed are the components of knowledge necessary to make a good judgement, the impact of our own prejudices in relation to projects or for that matter the cultural elements within an organisation which impact upon the decision making process. Elaine created a system to break this down and give reasons and logic to both the process and the human interaction necessary to improve the chances of success. Adopting her recommendations will improve teamwork and outcomes for your company.

Edward Roderick Hon LLD, Former CEO Christian Salvesen Plc

TAME, MESSY AND WICKED RISK LEADERSHIP, DAVID HANCOCK

This book takes project risk management firmly onto a higher and wider plane. We thought we knew what project risk management was and what it could do. David Hancock shows us a great deal more of both. David Hancock has probably read more about risk management than almost anybody else, he has almost certainly thought about it as much as anybody else and he has quite certainly learnt from doing it on very difficult projects as much as anybody else. His book draws fully on all three components. For a book which tackles a complex subject with breadth, insight and novelty – its remarkable that it is also a really good read. I could go on!

Dr Martin Barnes CBE FREng, President,
The Association for Project Management

This compact and thought provoking description of risk management will be useful to anybody with responsibilities for projects, programmes or businesses. It hits the nail on the head in so many ways, for example by pointing out

that risk management can easily drift into a check-list mindset, driven by the production of registers of numerous occurrences characterised by the Risk = Probablity x Consequence equation. David Hancock points out that real life is much more complicated, with the heart of the problem lying in people, so that real life resembles poker rather than roulette. He also points out that while the important thing is to solve the right problem, many real life issues cannot be readily described in a definitive statement of the problem. There are often interrelated individual problems with surrounding social issues and he describes these real life situations as 'Wicked Messes'. Unusual terminology, but definitely worth the read, as much for the overall problem description as for the recommended strategies for getting to grips with real life risk management. I have no hesitation in recommending this book.

Sir Robert Walmsley KCB FREng,
Chairman of the Board of the Major Projects Association

In highlighting the complexity of many of today's problems and defining them as tame, messy or wicked, David Hancock brings a new perspective to the risk issues that we currently face. He challenges risk managers, and particularly those involved in project risk management, to take a much broader approach to the assessment of risk and consider the social, political and behavioural dimensions of each problem, as well as the scientific and engineering aspects with which they are most comfortable. In this way, risks will be viewed more holistically and managed more effectively than at present.

Dr Lynn T Drennan, Chief Executive, Alarm,
the Public Risk Management Association

SUSTAINABILITY IN PROJECT MANAGEMENT, GILBERT SILVIUS, JASPER VAN DEN BRINK, RON SCHIPPER, ADRI KÖHLER AND JULIA PLANKO

Sustainability in Project Management thinking and techniques is still in its relatively early days. By the end of this decade it will probably be universal, ubiquitous, fully integrated and expected. This book will be a most valuable guide on this journey for all those interested in the future of projects and how they are managed in a world in peril.

Tom Taylor dashdot and vice-President of APM

Project Managers are faced with lots of intersections. The intersection of projects and risk, projects and people, projects and constraints... Sustainability in Projects and Project Management is a compelling, in-depth treatment of a most important intersection: the intersection of project management and sustainability. With detailed background building to practical checklists and a call to action, this book is a must-read for anyone interested in truly implementing sustainability, project manager or not.

Rich Maltzman, PMP, Co-Founder, EarthPM, LLC,
and co-author of *Green Project Management*,
Cleland Literature Award Winner of 2011

Great book! Based on a thorough review on existing relevant models and concepts the authors provide guidance for different stakeholders such as Project Managers and Project Office Managers to consider sustainability principles on projects. The book gets you started on sustainability in project context!

Martina Huemann, WU-Vienna University of Economics
and Business, Vienna Austria

While sustainability and green business have been around a while, this book is truly a 'call to action' to help the project manager, or for that matter, anyone, seize the day and understand sustainability from a project perspective. This book gives real and practical suggestions as to how to fill the sustainability/project gap within your organization. I particularly liked the relationship between sustainability and 'professionalism and ethics', a connection that needs to be kept in the forefront.

David Shirley, PMP, Co-Founder, EarthPM, LLC,
and co-author of *Green Project Management*,
Cleland Literature Award Winner of 2011

It is high time that quality corporate citizenship takes its place outside the corporate board room. This excellent work, which places the effort needed to secure sustainability for everything we do right where the rubber hits the road – our projects – has been long overdue. Thank you Gilbert, Jasper, Ron, Adri and Julia for doing just that! I salute you.

Jaycee Krüger, member of ISO/TC258
a technical committee for the creation of standards in Project,
Program and Portfolio Management, and chair of SABS/TC258,
the South African mirror committee of ISO/TC258

Sustainability is no passing fad. It is the moral obligation that we all face in ensuring the future of human generations to come. The need to show stewardship and act as sustainability change agents has never been greater. As project managers we are at the forefront of influencing the direction of our projects and our organisations. Sustainability in Project Management offers illuminating insights into the concept of sustainability and its application to project management. It is a must read for any modern project manager.

Dr Neveen Moussa, Project Manager, Adjunct Professor
of Project Management and past president of the Australian
Institute of Project Management

About the Series Editor

Professor Darren Dalcher is founder and Director of the National Centre for Project Management, a Professor of Project Management at the University of Hertfordshire and Visiting Professor of Computer Science at the University of Iceland.

Following industrial and consultancy experience in managing IT projects, Professor Dalcher gained his PhD from King's College, University of London. In 1992, he founded and chaired of the Forensics Working Group of the IEEE Technical Committee on the Engineering of Computer-Based Systems, an international group of academic and industrial participants formed to share information and develop expertise in project and system failure and recovery.

He is active in numerous international committees, standards bodies, steering groups, and editorial boards. He is heavily involved in organising international conferences, and has delivered many international keynote addresses and tutorials. He has written over 150 refereed papers and book chapters on project management and software engineering. He is Editor-in-Chief of the *International Journal of Software Maintenance and Evolution*, and of the *Journal of Software: Evolution and Process*. He is the editor of a major new book series, Advances in Project Management, published by Gower Publishing which synthesises leading edge knowledge, skills, insights and reflections in project and programme management and of a new companion series, Fundamentals of Project Management, which provides the essential grounding in key areas of project management.

He has built a reputation as leader and innovator in the area of practice-based education and reflection in project management and has worked with many major industrial, commercial and charitable organisations and government bodies. In 2008 he was named by the Association for Project Management as one of the top 10 influential experts in project management and has also been voted *Project Magazine's* Academic of the Year for his contribution in 'integrating and weaving academic work with practice'. He has been chairman of the APM Project Management Conference since 2009, setting consecutive attendance records and bringing together the most influential speakers.

He received international recognition in 2009 with appointment as a member of the PMForum International Academic Advisory Council, which features leading academics from some of the world's top universities and academic institutions. The Council showcases accomplished researchers, influential educators shaping the next generation of project managers and recognised authorities on modern project management. In October 2011 he was awarded a prestigious Honorary Fellowship from the Association for Project Management for outstanding contribution to project management.

He has delivered lectures and courses in many international institutions, including King's College London, Cranfield Business School, ESC Lille, Iceland University, University of Southern Denmark, and George Washington University. His research interests include project success and failure; maturity and capability; ethics; process improvement; agile project management; systems and software engineering; project benchmarking; risk management; decision making; chaos and complexity; project leadership; change management; knowledge management; evidence-based and reflective practice.

Professor Dalcher is an Honorary Fellow of the Association for Project Management, a Chartered Fellow of the British Computer Society, a Fellow of the Chartered Management Institute, and the Royal Society of Arts, and a Member of the Project Management Institute, the Academy of Management, the Institute for Electrical and Electronics Engineers, and the Association for Computing Machinery. He is a Chartered IT Practitioner. He is a Member of the PMI Advisory Board responsible for the prestigious David I. Cleland Project Management Award; of the APM Group Ethics and Standards Governance Board, and, until recently; of the APM Professional Development Board. He is a member of the OGC's International Reference Group for Managing Successful Programmes; and Academic and Editorial Advisory Council Member for *PM World Journal*, for which he also writes a regular column featuring advances in research and practice in project management.

National Centre for Project Management
University of Hertfordshire
MacLaurin Building
4 Bishops Square
Hatfield, Herts, AL10 9NE
Email: ncpm@herts.ac.uk